Strategic Voting

Synthesis Lectures on Artificial Intelligence and Machine Learning

Editors

Strategic Voting
Reshef Meir
2018

Predicting Human Decision-Making: From Prediction to Action
Ariel Rosenfeld and Sarit Kraus
2018

Game Theory for Data Science: Eliciting Truthful Information
Boi Faltings and Goran Radanovic
2017

Multi-Objective Decision Making
Diederik M. Roijers and Shimon Whiteson
2017

Lifelong Machine Learning
Zhiyuan Chen and Bing Liu
2016

Statistical Relational Artificial Intelligence: Logic, Probability, and Computation
Luc De Raedt, Kristian Kersting, Sriraam Natarajan, and David Poole
2016

Representing and Reasoning with Qualitative Preferences: Tools and Applications
Ganesh Ram Santhanam, Samik Basu, and Vasant Honavar
2016

Planning with Markov Decision Processes: An AI Perspective
Mausam and Andrey Kolobov
2012

Active Learning
Burr Settles
2012

Computational Aspects of Cooperative Game Theory
Georgios Chalkiadakis, Edith Elkind, and Michael Wooldridge
2011

Representations and Techniques for 3D Object Recognition and Scene Interpretation
Derek Hoiem and Silvio Savarese
2011

A Short Introduction to Preferences: Between Artificial Intelligence and Social Choice
Francesca Rossi, Kristen Brent Venable, and Toby Walsh
2011

Human Computation
Edith Law and Luis von Ahn
2011

Trading Agents
Michael P. Wellman
2011

Visual Object Recognition
Kristen Grauman and Bastian Leibe
2011

Learning with Support Vector Machines
Colin Campbell and Yiming Ying
2011

Algorithms for Reinforcement Learning
Csaba Szepesvári
2010

Data Integration: The Relational Logic Approach
Michael Genesereth
2010

Markov Logic: An Interface Layer for Artificial Intelligence
Pedro Domingos and Daniel Lowd
2009

Introduction to Semi-Supervised Learning
XiaojinZhu and Andrew B.Goldberg
2009

Action Programming Languages
Michael Thielscher
2008

Representation Discovery using Harmonic Analysis
Sridhar Mahadevan
2008

Essentials of Game Theory: A Concise Multidisciplinary Introduction
Kevin Leyton-Brown and Yoav Shoham
2008

A Concise Introduction to Multiagent Systems and Distributed Artificial Intelligence
Nikos Vlassis
2007

Intelligent Autonomous Robotics: A Robot Soccer Case Study
Peter Stone
2007

Strategic Voting

Reshef Meir

ISBN: 978-3-031-00451-3 paperback
ISBN: 978-3-031-01579-3 ebook
ISBN: 978-3-031-00024-9 hardcover

DOI 10.1007/978-3-031-01579-3

A Publication in the Springer series
SYNTHESIS LECTURES ON ARTIFICIAL INTELLIGENCE AND MACHINE LEARNING

Lecture #38
Series Editors: Ronald J. Brachman, *Jacobs Technion-Cornell Institute at Cornell Tech*
Peter Stone, *University of Texas at Austin*

Series ISSN
Print 1939-4608 Electronic 1939-4616

Strategic Voting

Reshef Meir
Technion–Israel Institute of Technology

SYNTHESIS LECTURES ON ARTIFICIAL INTELLIGENCE AND MACHINE LEARNING #38

ABSTRACT

Social choice theory deals with aggregating the preferences of multiple individuals regarding several available alternatives, a situation colloquially known as *voting*.

There are many different voting rules in use and even more in the literature, owing to the various considerations such an aggregation method should take into account. The analysis of voting scenarios becomes particularly challenging in the presence of strategic voters, that is, voters that misreport their true preferences in an attempt to obtain a more favorable outcome. In a world that is tightly connected by the Internet, where multiple groups with complex incentives make frequent joint decisions, the interest in strategic voting exceeds the scope of political science and is a focus of research in economics, game theory, sociology, mathematics, and computer science.

The book has two parts. The first part asks "are there voting rules that are truthful?" in the sense that all voters have an incentive to report their true preferences. The seminal Gibbard-Satterthwaite theorem excludes the existence of such voting rules under certain requirements. From this starting point, we survey both extensions of the theorem and various conditions under which truthful voting is made possible (such as restricted preference domains). We also explore the connections with other problems of mechanism design such as locating a facility that serves multiple users.

In the second part, we ask "what would be the outcome when voters do vote strategically?" rather than trying to prevent such behavior. We overview various game-theoretic models and equilibrium concepts from the literature, demonstrate how they apply to voting games, and discuss their implications on social welfare.

We conclude with a brief survey of empirical and experimental findings that could play a key role in future development of game theoretic voting models.

KEYWORDS

social choice, game theory, strategic voting, mechanism design, implementation

To Yaara and Arad.

Contents

Preface

In a typical voting scenario, a group of voters needs to collectively choose one out of several alternatives based on their private preferences. Examples include a committee that selects from a pool of candidates for a faculty position or an academic award, countries in an international forum voting on the adoption of a new environmental treaty, and automated agents that vote on a preferred meeting time on behalf of their users. As the satisfaction of each voter is determined by the selected alternative, which is in turn affected by the actions (namely, the ballots) of others, casting a vote is in fact playing a strategic game.

It was the American economist Anthony Downs who argued that political decisions, and voting behavior in particular, could be analyzed like other economic interactions. Downs' influential paper *An Economic Theory of Political Action in a Democracy* [Downs, 1957] has led to a barrage of game theoretic models for strategic voting in the 1970s and 1980s. With the rise of the Internet, the automated aggregation of individual preferences has become both frequent and large scale, and the new algorithmic challenges led to the emergence of *computational social choice* as a field of academic research.

The study of strategic voting is an effort to utilize game theory, which attempts to model and predict rational behavior in a wide range of economic and social interactions, to explain the strategic decisions of voters. Hence, this book is built on two firm pillars, namely *social choice theory* and *game theory*. In multiple places throughout the book, I highlight the connections between these areas and demonstrate how definitions from social choice can be "translated" to a game theoretic language (and vice versa). I also put emphasis on a computational approach, highlighting recent conceptual and technical contributions by computer scientists.[1]

The most fundamental concept of game theory is that of *equilibrium*: a state in which no agent wants to change her action. Much of the game theory literature revolves around characterizing the equilibria of various games and studying their properties. More recently, mechanism design (sometimes called "reversed game theory") entered the scene, with the purpose of constructing games where participants have an incentive to play honestly, or where strategic behavior leads to a desired equilibrium outcome. Interestingly, while voting scenarios can be very naturally described as games, a naïve attempt to apply either of the mechanism design approaches above fails spectacularly: the *Gibbard-Satterthwaite theorem* states that no honest voting system could be designed even under basic requirements, and applying the most important equilibrium concept—Nash equilibrium—to common voting rules is often uninformative.

This book is the story of the main approaches attempted so far in overcoming these two setbacks: the search for truthful voting rules (Part I); and the alternatives to Nash equilibrium

[1]At least two of the recent theorems mentioned in the book were proven with the nontrivial aid of a computer!

that can be used to better understand the outcome of strategic voting (Part II). Since much of the literature in the different communities that considered these questions remained separate, my purpose here is to overview existing models and results in a way that makes them comparable across fields and disciplines.

WHO IS THIS BOOK FOR?

The readers who will benefit the most from this book are those who already feel comfortable with either game theory or social choice but lack solid background in the other. The current book provides a very brief introduction of the main concepts in each of these fields. There are many introductory textbooks for game theory. Of which, the current book tries to be consistent with the definitions and terminology of Leyton-Brown and Shoham [2008]. Similarly, the reader will find in Zwicker [2016] all the necessary background on social choice theory.

Researchers in the area of strategic voting who are experts in some of the topics in this book may use the book to expand their horizon in the other topics.

The book may also be beneficial as a teaching aid in advanced courses on social choice and/or game theory, and each chapter contains several exercises at a graduate student level.

Some familiarity with basic combinatorics, algorithms, and computational complexity is also assumed, roughly at the level of a computer science undergraduate student.

WHAT IS NOT IN THIS BOOK?

Strategic voting is a very broad area and naturally this book cannot cover everything. First, the book does not cover scenarios that are often considered part of social choice, such as tournaments, matching and fair division, all of which and more are covered in the *Handbook of Computational Social Choice* [Brandt et al., 2016].

Second, political elections involve many factors which are less relevant for the study of general preference aggregation. For example, it is possible to think of strategic behavior by the candidates, election chairs, external parties, and so on. These are not considered in the book, and neither are changing preferences or complex incentives that depend on social utilities. This book focuses almost exclusively on voters with fixed straightforward preferences who try to get their preferred candidate to win.

Lastly, some of the theorems and definitions are presented in a simplified form, so as to focus on a special case of interest, and many of the proofs are omitted.

Reshef Meir
April 2018

Acknowledgments

This book is the fruit of many discussions, conversations, e-mail correspondences, and sometimes arguments, with numerous people from the game theory and social choice communities: Yakov Babichenko, Gal Cohensius, Edith Elkind, Roy Fairstein, Piotr Faliszewski, Kobi Gal, Svetlana Obraztsova, Hongyao Ma, Dov Monderer, Matías Núñez, David Parkes, Maria Polukarov, Ariel D. Procaccia, Jeffrey S. Rosenschein, Moshe Tennenholtz, Anaelle Wilczynski, and many others.

In particular, Itai Arieli, Umberto Grandi, Ron Lavi, Barton Lee, Omer Lev, Ilan Nehama, Aris Filos-Ratsikas, Arkadii Slinko, Zoi Terzopoulou, Alan Tsang, and Mark Wilson have been kind enough to send comments on the draft.

This book would not be possible without the initiative, encouragement and guidance of Morgan & Claypool and in particular of Mike Morgan, Peter Stone, Christine Kiilerich, and Clovis L. Tondo.

Special thanks go to Haris Aziz, Ulle Endriss, and Jérôme Lang who have carefully read drafts of this book and helped to improve it in many ways.

The mistakes are all my own.

The author is partly supported by the Israel Science Foundation (ISF) under Grant No. 773/16.

Reshef Meir
April 2018

CHAPTER 1

Introduction

It is a big day today at Jane's preschool. The last snow just melted away and the teacher suggested the kids decorate the playground and turn it into an exciting adventure park. However, artistic disagreements arose over what would be the best decoration theme. The teacher has decided to settle the issue by asking kids to vote over the four most popular ideas, thereby also providing them with a valuable lesson on the importance of democracy and compromise.

Each child received a single ballot, which they could attach to one of the four alternatives presented to them on the board: a Firehouse, a Magic Treehouse, a Castle, or a Zoo (henceforth, F, MT, C, and Z, respectively).[1] Jane's favorite theme is Z, followed by MT, C and F, and she is excited about adding her vote to the Zoo option (see Figure 1.1). However, by the time she gets to the board she sees the following scores:

F	MT	C	Z
5	2	5	1

As Jane is the last child who has not yet voted, she realizes with disappointment that Z will not win even with her support. Instead, Jane votes for C, securing the selection of C, which she still prefers over F.

The above behavior is an example of a *strategic manipulation*. In this case, Jane changed the outcome by misreporting her true preferences. We should keep in mind that it is quite possible that the other kids also had applied similar strategic reasoning, and may even have speculated on how Jane might vote prior to casting their own vote. Therefore, the seemingly naïve situation described here is in fact a *game*, whose outcome depends not only on the *preferences* of the participants, but also on their *strategic behavior*.

In contrast, suppose that the kids only had two options to vote for (say, F and MT). In this case, regardless of how other children had voted, Jane would always be better off voting for MT, which she likes better. In other words, when there are only two alternatives (and we choose the one with more votes), then the voting procedure is *truthful*.

Chapter 2 of the book lays out useful definitions from social choice and game theory that let us formalize scenarios like the above example. The rest of the book is divided into two parts that ask different questions.

[1]The author is still unsure as to what exactly is a "Firehouse."

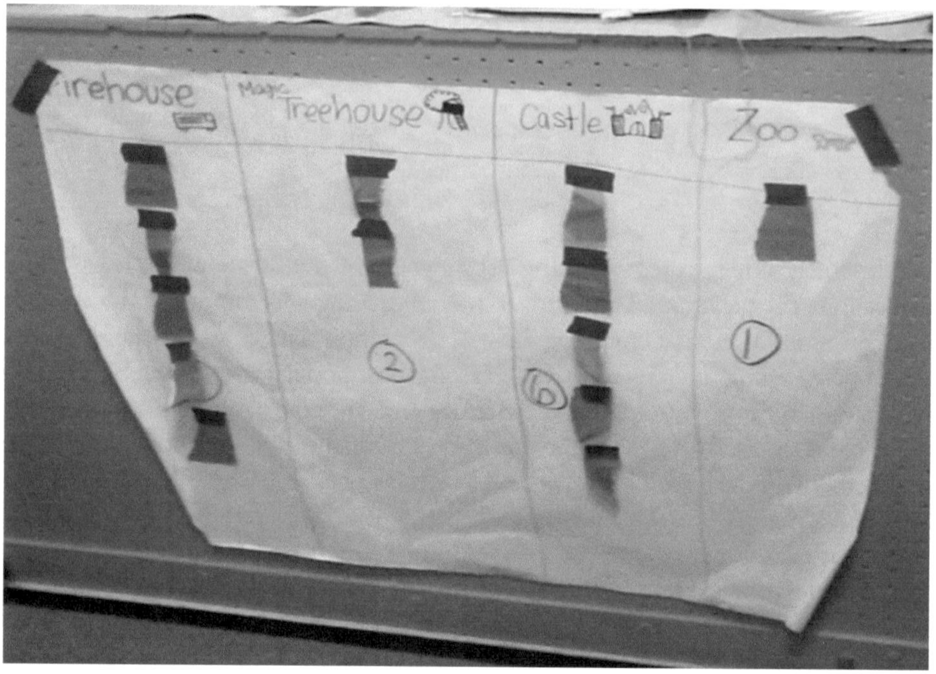

Figure 1.1: A snapshot of the voting process in Botanic Gardens' preschool.

PART I: THE QUEST FOR TRUTHFUL VOTING

The first part of the book follows the common approach that asserts that voting manipulations distort the will of the society and should thus be avoided, if possible.[2]

We begin with a fundamental negative result that had shaped the field of strategic voting, namely the Gibbard-Satterthwaite (G-S) theorem. The theorem proves that when there are at least three alternatives, the only onto voting rules where voting manipulations do not exist are *dictatorial* (Chapter 3). We then provide some follow-up results showing that manipulations are frequent, and even more so when considering manipulations by groups of voters. These results highlight the inevitability of strategic behavior in voting.

Chapter 4 overviews the main approaches that were suggested to overcome the G-S theorem and regain truthfulness in voting:

- restricting the domain, so that not all preferences are possible;
- designing voting rules where strategic manipulations exist, but are computationally difficult to find by a voter with limited resources;

[2]Suppose for example that most of the kids in our example preferred Z, yet voted for other options due to strategic reasoning!

- allowing voting rules that use randomization; and
- relaxing the truthfulness requirement so as to allow voting rules that are "almost truthful."

Chapter 5 ties the search for truthful voting rules with the more recent and general agenda of *mechanism design*. In this chapter we assume there is a quantifiable social goal (sum of agents' utilities) that we want to maximize and consider how close we can get to it using truthful voting rules. This turns out to be a difficult problem, but one that can be efficiently overcome once money is introduced, using the Vickrey–Clarke–Groves (VCG) mechanism. Focusing again on mechanisms without money, we explore the connections between voting (which deals with aggregating rankings); facility location (which deals with aggregating locations in a metric space); and aggregation of judgments over some logical agenda. Interestingly, judgment aggregation can be thought of both as a special case of facility location and as a generalization of voting. Yet, defining strategic behavior in judgment aggregation raises new conceptual challenges.

The results surveyed in Part I can be roughly divided into two types: *negative results* (like the G-S theorem) showing that truthful voting is impossible if certain requirements are made, and *positive results* showing the existence of voting rules and mechanisms with some desirable properties.

PART II: VOTING EQUILIBRIUM MODELS

The second part of the book forgoes the attempt to extract voters' true preferences. Instead, we accept that voters will behave strategically and apply game theoretic equilibrium analysis. For example, given the full preferences of all children in the above example and fixing the voting rule, we could ask what would be the likely outcome.

Most of the results described in this part are not "positive" or "negative" but typically aim to understand the properties of equilibrium outcomes under various voting rules and behavioral assumptions.

Our basic observation is that the most important and widely known game theoretic solution concept—Nash equilibrium—is uninformative in almost all voting scenarios. This is because almost any voting profile in common voting rules is a Nash equilibrium regardless of voters' preferences due to the fact that a single voter can rarely influence the outcome. In particular a single voter rarely has an incentive to vote strategically (or to vote at all). This simple observation had triggered the suggestion of many theoretical models that justify various forms of strategic voting.

To facilitate the understanding of the presented models we include two running examples that are used throughout Part II. We also introduce desiderata by which various models can be compared and evaluated.

Chapter 6 reviews several equilibrium models under the assumption that voters are rational and participate in a one-shot voting game. In particular, we consider models that emphasize the role of voters' *uncertainty* on the actions of others, or on the voting process itself such as the probability that votes are miscounted.[3] In the presence of uncertainty, a voter never knows *for sure* whether she is pivotal or not, which introduces more complex incentives.

Chapter 7 introduces game dynamics, that is, when voters vote one-by-one or when the voting process allows the nomination of a new candidate in every round. For example, we could imagine that in the playground example above, children would be allowed to change their vote after observing the votes of others. The main question studied in this chapter is under what conditions the voters converge to an equilibrium.

In Chapter 8 equilibrium and convergence analysis continue, but the standard game-theoretic assumptions are replaced with decision making models that aim to better capture real voting behavior. We make a distinction between "ad hoc" voting heuristics that apply a simple rule of thumb and heuristics that are based on the explicit but uncertain beliefs the voter has on the actions of others.

At the end of every chapter the reader will find a list of exercises that demonstrate key theoretical ideas. The final chapter concludes with a brief overview of empirical and experimental findings on strategic voting by people, and suggests some future directions for researchers in the field.

A NOTE ON RELATED LITERATURE

The book aims to at least "touch" all major topics and approaches in strategic voting, but the space and depth dedicated to cover each topic ranges considerably. For example, implementation theory is a mature topic that can fill several books on its own and was squeezed into a single section, whereas the relatively new topic of iterative voting was given a whole chapter. These choices are purely subjective and reflect the author's areas of interest and expertise. References for more extensive survey articles and relevant chapters of textbooks are provided where it seems fit.

[3]The keen-eyed reader may have noticed that the five votes for C in Figure 1.1 have been counted as six. This of course may also happen in large scale voting.

CHAPTER 2

Basic Notation

We generally denote sets by uppercase letters (e.g., $A = \{a_1, a_2, \ldots\}$) and vectors by bold letters (e.g., $\boldsymbol{a} = (a_1, a_2, \ldots)$). The notation $\boldsymbol{a}_{-i} \triangleq (a_1, \ldots, a_{i-1}, a_{i+1}, \ldots)$ stands for all entries of $\boldsymbol{a}$ except a_i. Similarly, for a subset of indices I, $\boldsymbol{a}_{-I}$ includes all entries except $a_i, i \in I$.

For a finite set X, we denote by $\mathcal{R}(X)$ the set of all weak orders over X, and by $\mathcal{L}(X)$ the set of all linear (strict) orders over X. For $L \in \mathcal{L}(X)$, denote by top(L) the first element of L. We also denote by $\mathcal{U}(X)$ the set of all functions of the form $U : X \to \mathbb{R}$. For an integer k, we denote $[k] \triangleq \{1, \ldots, k\}$ and $[k]_0 \triangleq \{0, 1, \ldots, k\}$.

2.1 SOCIAL CHOICE

We follow the notation and definitions in the introductory chapter of *The Handbook of Computational Social Choice* [Zwicker, 2016] as much as possible.

A voting scenario is defined by a set of m *candidates*, or *alternatives*, A, a set of n voters N, and a *preference profile* $\boldsymbol{R} = (R_1, \ldots, R_n)$, where each $R_i \in \mathcal{R}(A)$. For $a, b \in A, i \in N$, candidate a precedes b in R_i (denoted $a \succ_i b$) if voter i prefers candidate a over candidate b. An indifference relation is denoted by $a \sim_i b$, and weak preference by $a \succsim_i b$.

In most sections we will assume that preferences are strict, where a strict preference profile is denoted by $\boldsymbol{L} = (L_1, \ldots, L_n) \in \mathcal{L}(A)^n$. For $c \in A$, $L_i^{-1}(c) \in [m]$ is the rank of candidate c in L_i; top$(L_i) \in A$ is i's most preferred candidate (i.e., $L_i^{-1}(\text{top}(L_i)) = 1$). Given a strict preference order L and a subset of candidates $B \subseteq A$, $L|_B$ is the (unique) preference order on B that agrees with L. For a profile $\boldsymbol{L} = (L_i)_{i \in N}$, we define $\boldsymbol{L}|_B \triangleq (L_i|_B)_{i \in N} \in \mathcal{L}(B)^n$.

Definition 2.1 (Social choice correspondence). A *social choice correspondence* (SCC) is a function $F : \mathcal{L}(A)^n \to 2^A \setminus \{\emptyset\}$, that is, accepts a preference profile $\boldsymbol{L}$ as input, and outputs a nonempty set of winning candidates.

Definition 2.2 (Social choice function). An SCC F is *resolute* if $|F(\boldsymbol{L})| = 1$ for all $\boldsymbol{L}$. Resolute SCCs are also called *social choice functions* (SCF). We typically denote SCFs by a lower case letter f.

The definitions above assume that the social choice correspondence/function is *deterministic*. Later in the book we also define randomized voting rules.

The *range* of an SCF f is the set $\text{range}(f) \triangleq \{f(\boldsymbol{L}) : \boldsymbol{L} \in \mathcal{L}(A)^n\}$. An SCF f is *onto* if $\text{range}(f) = A$. An SCF f is *unanimous* if for all $L \in \mathcal{L}(A)$, $f(L, L, \ldots, L) = \text{top}(L)$. Clearly any unanimous voting rule is also onto.

Definition 2.3 (Social welfare function). A *social welfare function* (SWF) is a function $\hat{f} : \mathcal{L}(A)^n \to \mathcal{R}(A)$, that is, accepts a preference profile $\boldsymbol{L}$ as input, and outputs a weak order over candidates. A SWF that always returns a strict order is called *resolute*.

SCCs, SCFs, and SCWs can be naturally extended to functions of weak preference profiles. Also note that every resolute SCW $\hat{f}$ induces a unique SCF f, where $f(\boldsymbol{L}) = \text{top}(\hat{f}(\boldsymbol{L}))$ for all $\boldsymbol{L}$. Our focus in this book will be on SCFs, and we will reserve the term *voting rule* for a SCF unless stated otherwise.

Common voting rules Some common voting rules that are mentioned in this book are based on computing some score $s(c, \boldsymbol{L})$, and selecting the candidate with the highest score as the winner as follows.

Plurality: The winner is the candidate ranked first by the largest number of voters. Formally, $s(c, \boldsymbol{L}) = |\{i \in N : \text{top}(L_i) = c\}|$.

Veto: The winner is the candidate ranked last by the smallest number of voters. Formally, $s(c, \boldsymbol{L}) = |\{i \in N : L_i(m) \neq c\}|$.

Positional Scoring Rules (PSR): Given a non-decreasing scoring vector $\boldsymbol{\alpha} = (\alpha_1, \ldots, \alpha_m)$ with $\alpha_1 > \alpha_m$, a candidate gets a score of α_1 for each time she is ranked first, α_2 for each time she is ranked second, and so on. Formally, the score of each candidate c is $s(c, \boldsymbol{L}) = \sum_{i \in N} \alpha_{L_i^{-1}(c)}$.

Borda: A special case of a PSR with $\alpha = (m-1, m-2, \ldots, 0)$.

k-Approval: A special case of a PSR where the first k entries of α are 1 and the rest are 0. Note that 1-Approval is Plurality and $(m-1)$-Approval is Veto.

Note that all of the voting rules above are unanimous, so if there is a complete consensus on the winning alternative it must be elected. A weaker form of consensus is a *Condorcet winner*, which only has to have the support of a majority of voters vs. any other candidate in a pairwise match.[1] Formally, denote by $p(c, c', \boldsymbol{L}) \triangleq |\{i \in N : c \succ_i c'\}| - \frac{n}{2}$ the number of voters who prefer c over c' (shifted so that $p(c, c', \boldsymbol{L}) + p(c', c, \boldsymbol{L}) = 0$).

Definition 2.4 (Condorcet winner). An alternative $c \in A$ is a (strict) *Condorcet winner* in profile $\boldsymbol{L}$ if $p(c, c', \boldsymbol{L}) > 0$ for all $c' \in A \setminus \{c\}$. $c \in A$ is a *weak Condorcet winner* if the above holds with a weak inequality.

[1]For other forms of consensus and their use to study and define voting rules, see Elkind and Slinko [2016].

Similarly, an alternative c is a [strict/weak] *Condorcet loser* if all other alternatives beat c. A Condorcet winner does not always exist, and there can be at most one strict Condorcet winner. A voting rule is said to be *Condorcet consistent* if it always selects a weak Condorcet winner when one exists. The following rules are Condorcet consistent.

Copeland: The score of a candidate is the number of other candidates that she beats in a pairwise vote. Formally, $s(c, \boldsymbol{L}) = |\{c' : p(c, c', \boldsymbol{L}) > 0\}|$.[2]

Maximin: The score of a candidate is the worst margin she has against any other candidate. Formally, $s(c, \boldsymbol{L}) = \min_{c' \neq c} p(c', c, \boldsymbol{L})$.

Note that strictly speaking, all the voting rules described above are SCCs as there may be ties.[3] However any SCC with an additional tie-breaking rule is an SCF. We assume lexicographic tie-breaking throughout the book unless specified otherwise. This means that there is a fixed linear order over A that does not depend on voters' preferences or actions. In case of a tie, the tied candidate that appears first in this order is the winner.

Plurality and Veto are also PSRs with $\alpha = (1, 0, 0, \ldots, 0)$ and $\alpha = (1, 1, \ldots, 1, 0)$, respectively. Two other common rules are as follows.

Plurality with Runoff: The winner is the more preferred candidate among the two candidates with the highest Plurality score.

Single Transferable Vote (STV): Each round we remove the candidate with the lowest Plurality score until only one candidate remains.

Throughout the book we will use the functions f^{PL}, f^{VL}, and f^{BL} to denote the Plurality, Veto, and Borda rules with lexicographic tie-breaking.

More properties We next define several very basic properties (sometimes called *axioms*) that a voting rule may either satisfy or violate. Recall we have already defined the properties of *onto* and *unanimity*.

Definition 2.5 (Anonymity). An SCC F is *anonymous* if $F(\pi(\boldsymbol{L})) = F(\boldsymbol{L})$ for any permutation $\pi : N \to N$ of the voters.

Definition 2.6 (Neutrality). An SCC F is *neutral* if $F(\pi(\boldsymbol{L})) = \pi(F(\boldsymbol{L}))$ for any permutation $\pi : A \to A$ of the alternatives.

[2] There are variations for the score candidates should get in case of a tie. In this version both candidates get 0.
[3] In fact, since all these rules are based on a numerical score, they can each be interpreted also as SWFs.

All common voting rules mentioned above are anonymous and neutral when considered as SCCs, but any deterministic tie-breaking rule will violate at least one of them.

Definition 2.7 (Pareto optimal candidate). A candidate $c \in A$ is *Pareto optimal* in profile $\boldsymbol{L}$ if for any other candidate c', there is at least one player i s.t. $c \succ_i c'$.

Definition 2.8 (Pareto voting rule). A voting rule f is *Pareto* if for any profile $\boldsymbol{L}$, $f(\boldsymbol{L})$ is Pareto optimal in $\boldsymbol{L}$.

All common voting rules mentioned above are Pareto, regardless of tie-breaking.

2.2 GAME THEORY

We will use in this book some basic concepts from the theory of non-cooperative games. We mainly follow the notations and definitions from Leyton-Brown and Shoham [2008].

Definition 2.9 (Game). A (finite, n-person, non-cooperative) *game* is a tuple $\langle N, \mathcal{A}, \boldsymbol{u} \rangle$, where:

- N is a finite set of n players, indexed by i;
- $\mathcal{A} = A_1 \times \cdots \times A_n$, where A_i is a finite set of actions available to player i. Each vector $\boldsymbol{a} = (a_1, \ldots, a_n) \in \mathcal{A}$ is called an *action profile*; and
- $\boldsymbol{u} = (u_1, \ldots, u_n)$ where $u_i : \mathcal{A} \to \mathbb{R}$ is a real-valued utility (or payoff) function for player i.

For example, the famous *prisoner's dilemma* game is described in Figure 2.1: The set of players is $N = \{1, 2\}$, where 1 selects a row and 2 selects a column; $A_1 = A_2 = \{C, D\}$ (which stand for the actions "Cooperate" and "Defect"); and the numbers in each cell represent the utility for player 1 and 2, respectively (e.g., $u_2(C, D)) = 4$).

	C	D
C	$3,3$	$0,4$
D	$4,0$	$1,1$

	C	D
C	a	b
D	c	d

Figure 2.1: On the left, one variation of the prisoner's dilemma game. On the right, a 2×2 game form.

Definition 2.10 (Better reply). An action a'_i is a *(weakly) better reply* to the action profile $\boldsymbol{a}$ if $u_i(a'_i, \boldsymbol{a}_{-i})$ is (weakly) greater than $u_i(a_i, \boldsymbol{a}_{-i})$.

Definition 2.11 (Best reply). Player i's *best reply* to the action profile $\boldsymbol{a}_{-i}$ is an action $a_i^* \in A_i$ such that $u_i(a_i^*, \boldsymbol{a}_{-i}) \geq u_i(a_i, \boldsymbol{a}_{-i})$ for all actions $a_i \in A_i$.

Consider any profile $\boldsymbol{a}$, and an action a_i^* that is best reply to $\boldsymbol{a}_{-i}$:

- a_i^* is a weakly better reply to $\boldsymbol{a}$ and
- there is a (strict) better reply a_i' to $\boldsymbol{a}$ if and only if $a_i^* \neq a_i$.

By referring to a best reply to $\boldsymbol{a}$ we always mean a best reply to $\boldsymbol{a}_{-i}$.

Definition 2.12 (Pure Nash equilibrium). An action profile $\boldsymbol{a} = (a_1, \ldots, a_n)$ is a *pure Nash equilibrium* (PNE) if, for all agents i, a_i is a best reply to $\boldsymbol{a}_{-i}$. Equivalently, if no agent has a better reply to $\boldsymbol{a}$.

Definition 2.13 (Domination). Let a_i and a_i' be two actions of player i, and $\mathcal{A}_{-i}$ be the set of all action profiles of the remaining players. Then:

1. a_i *strictly dominates* a_i' if for all $\boldsymbol{a}_{-i} \in \mathcal{A}_{-i}$, it is the case that $u_i(a_i, \boldsymbol{a}_{-i}) > u_i(a_i', \boldsymbol{a}_{-i})$;
2. a_i *weakly dominates* a_i' if for all $\boldsymbol{a}_{-i} \in \mathcal{A}_{-i}$, it is the case that $u_i(a_i, \boldsymbol{a}_{-i}) \geq u_i(a_i', \boldsymbol{a}_{-i})$, and for at least one $\boldsymbol{a}_{-i} \in \mathcal{A}_{-i}$, it is the case that $u_i(a_i, \boldsymbol{a}_{-i}) > u_i(a_i', \boldsymbol{a}_{-i})$.

Definition 2.14 (Dominant strategy). A strategy is strictly (weakly) dominant for an agent if it strictly (weakly) dominates any other strategy for that agent.

We note that neither pure Nash equilibria nor dominant strategies are guaranteed to exist in a game.

Remark 2.15 The definitions above naturally extend to *mixed strategies*, which are randomized actions. For reasons that will become apparent later, most of the literature on strategic voting does not deal with mixed actions. Hence, the (simplified) definitions above will be sufficient for our needs in this book.

Game forms Informally, a game form specifies the actions and outcomes in a game, but without the utilities of the individual players. Thus in a sense it only requires the first two bullets of Definition 2.9. Since in principle there may be many action profiles leading to the same outcome, it will be useful to refer to action profiles and outcomes separately.

Definition 2.16 (Game form). A *game form* is a tuple $\langle N, \mathcal{A}, A, g \rangle$, where N and $\mathcal{A}$ are as in Definition 2.9; A is a finite set of outcomes (or alternatives); and $g : \mathcal{A} \to A$ is a function from action profiles to outcomes. We sometimes refer to such a game form simply as g, omitting the other parameters.

Definition 2.17 (Cardinal utility). A *cardinal utility function* for a set of alternatives A is a function $U \in \mathcal{U}(A)$. The cardinal utility function U is *strict* if there are no $a, b \in A$ with $U(a) = U(b)$.

We denote by $\mathcal{U}_L(A) \subseteq \mathcal{U}(A)$ all strict cardinal utility functions over the set A.

Definition 2.18 (Ordinal utility). An *ordinal utility function* for a set of alternatives A is a weak order $R \in \mathcal{R}(A)$.

Clearly, every cardinal utility function U induces a unique ordinal utility R where alternatives are sorted in non-decreasing order according to U. Similarly, a strict cardinal utility induces a unique linear preference L.

Definition 2.19 (Fit). For a given ordinal utility R and cardinal utility U, we say that R *fits* U if for all pairs $a, b \in A$, $U(a) \geq U(b)$ if and only if $a \succsim_R b$.

Any game form g (or in full notation, $\langle N, \mathcal{A}, A, g \rangle$) together with utility functions $\boldsymbol{U} = (U_i)_{i \in N}$ induce a unique game denoted by $\langle g, \boldsymbol{U} \rangle$. This is simply the normal form game $\langle N, \mathcal{A}, \boldsymbol{u} \rangle$, where $u_i(\boldsymbol{a}) \triangleq U_i(g(\boldsymbol{a}))$ for all $i \in N$ and $\boldsymbol{a} \in \mathcal{A}$. For example, consider the game form on Figure 2.1 (right). The game on the left is obtained by setting $U_1(a) = 3, U_2(a) = 3, U_1(b) = 0$ and so on.

Ordinal games We can take any game form and add ordinal rather than cardinal utilities, resulting in an *ordinal game*. Note that since we only consider pure actions, the definitions of best response, PNE, and dominance extend naturally to ordinal games. The set of PNEs in an ordinal game $\langle g, \boldsymbol{R} \rangle$ is the same as in any cardinal game $\langle g, \boldsymbol{U} \rangle$ such that U_i fits R_i for all $i \in N$.

Mechanisms have many variations and often assume particular structure of game forms and/or preferences. We thus avoid a general definition of a mechanism, and provide specific definitions where needed along the book.

Definition 2.20 (Pareto domination). An outcome $a \in A$ *Pareto dominates* $a' \in A$ in game $\langle g, \boldsymbol{R} \rangle$ if for all $i \in N$, $a \succsim_i a'$, and for at least one player $a \succ_i a'$.

Definition 2.21 (Pareto optimality). An outcome $a \in A$ in game $\langle g, \boldsymbol{R} \rangle$ is *Pareto optimal* (or *Pareto efficient*) if it is not Pareto dominated by any other outcome.

These definitions naturally extend to games with cardinal utilities.

g_1	a	b	c
a	a	a	a
b	b	b	b
c	c	c	c

g_2	a	b	c
a	a	a	a
b	a	b	b
c	a	b	c

g_3	x	y
a	a	b
b	b	c
c	c	a

g_4	x	y	z	w
a	ax	ay	az	aw
b	bx	by	bz	bw
c	cx	cy	cz	cw

Figure 2.2: Four examples of game forms with two agents. g_1 is a dictatorial game form with 3 candidates (the row agent is the dictator). g_2 is the Plurality voting rule with 3 candidates and lexicographic tie-breaking. f_3 and f_4 are non-standard game forms. In g_3, $A_1 = A = \{a, b, c\}, A_2 = \{x, y\}$. Note that g_4 is completely general (there are 3×4 possible outcomes in A, one for each voting profile) and can represent any 3-by-4 game.

$\langle g_2, \boldsymbol{L}^1 \rangle$	a	$\boldsymbol{b}$	c
$\boldsymbol{a}$	$3, 1*$	$\boldsymbol{3, 1*}$	$3, 1*$
b	$3, 1$	$2, 3$	$2, 3$
c	$3, 1$	$2, 3$	$1, 2$

$\langle g_2, \boldsymbol{L}^2 \rangle$	a	$\boldsymbol{b}$	c
a	$1, 2*$	$1, 2$	$1, 2$
b	$1, 2$	$2, 3*$	$2, 3$
$\boldsymbol{c}$	$1, 2$	$\boldsymbol{2, 3*}$	$3, 2$

Figure 2.3: Two games derived from the Plurality game form (g_2 in Figure 2.2). Consider first $\boldsymbol{L}^1 \triangleq (a \succ_1 b \succ_1 c, b \succ_2 c \succ_2 a)$. The ordinal utilities of both voters (higher is better) are on the left figure, and Nash equilibria are marked with *. The truthful actions and profile are marked in bold. We show the same in the right figure for the same game form with preferences $\boldsymbol{L}^2 \triangleq (c \succ_1 b \succ_1 a, b \succ_2 c \succ_2 a)$.

2.3 GAME FORMS ARE VOTING RULES

We next describe the tight connection between game forms and voting rules, which enables us to use them almost interchangeably.

Definition 2.22 (Standard game form). A game form g is *standard* if $A_i = \hat{A}$ for all i, and $\hat{A}$ is either $\mathcal{L}(A)$ (the set of permutations over A) or a coarsening of $\mathcal{L}(A)$.

Thus, a standard game form coincides with our definition of a voting rule, or SCF; see Figure 2.2 for examples. Each particular voting instance that includes voters' real preferences is an ordinal game, see Figure 2.3 for an example. Indeed, most common voting rules including Plurality, all PSRs, Maximin, STV, etc., are standard game forms once we fix a deterministic tie-breaking rule.

Approval is an example of a non-standard game form, where every voter may vote to any subset of candidates, adding one point to each (note that it does not match our definition of SCF or SCC). Another example is the *Direct Kingmaker* rule [Dutta, 1984], where voter n specifies a number $i \in \{1, \ldots, n-1\}$, and the winner is $\text{top}(L_i)$.

Nash equilibrium in voting Once we have a game, the common game theoretic approach is to analyze its equilibria.

Denote by $NE_g(\boldsymbol{L}) \subseteq A$ all candidates that win in some pure Nash equilibrium of the game $\langle g, \boldsymbol{L} \rangle$ (similarly for games with weak or cardinal preferences). Figure 2.3 shows two different games derived from the Plurality voting rule, and their pure Nash equilibria.

Consider Plurality voting with $n \geq 3$ voters. It is easy to see that any profile in which all voters vote for the same candidate is a Nash equilibrium. This is true even if this candidate is ranked last by all voters in $\boldsymbol{L}$, since no single voter can change the outcome. Therefore, $NE_{f^{PL}}(\boldsymbol{L}) = A$ for all $\boldsymbol{L}$. This already indicates that using Nash equilibrium as a way to predict or recommend voters' actions is problematic. Indeed, in all common voting rules, almost any voting profile is a Nash equilibrium regardless of preferences. In Chapter 6 we return to the notion of equilibrium in voting and consider refinements and variations that are more reasonable and more useful as a solution concept.

PART I

The Quest for Truthful Voting

CHAPTER 3

Strategyproofness and the Gibbard–Satterthwaite Theorem

3.1 VOTING MANIPULATIONS

Examples of a manipulation Recall the preschool voting example from the introduction. To present it in more formal terms, we would define a set of four alternatives $A = \{F, MT, C, Z\}$ and a preference profile $\boldsymbol{L}$ with $n = 14$ voters, of which five rank F at the top, two rank MT, and so on. In particular Jane would be one of the voters ranking Z at the top, with her full preferences being $L_j = Z \succ MT \succ C \succ F$. In profile $\boldsymbol{L}$, Jane (voter j) has a manipulation in the Plurality voting game $\langle f^{PL}, \boldsymbol{L} \rangle$, which is any vote L'_j that ranks C at the top (e.g., $L'_j = C \succ MT \succ Z \succ F$).

We next show an example of manipulation under the Borda rule f^{BL}. Consider an election with the following preference profile $\boldsymbol{L}$:

$$\begin{array}{l|l} L_1 & b \succ_1 a \succ_1 c \succ_1 d \\ L_2 & b \succ_2 a \succ_2 c \succ_2 d \\ L_3 & a \succ_3 b \succ_3 c \succ_3 d \end{array}$$

Candidate b is the winner, beating candidate a with 8 points to 7. However, voter 3 can manipulate the outcome by reporting $L'_3 = a \succ c \succ d \succ b$. Then b's score goes down to 6, and a (voter 3's favorite candidate) wins with 8 points.

A natural question is whether there are voting rules where such manipulations are impossible, that is, where a voter can never gain from lying about her preferences.

Strategyproofness and monotonicity

Definition 3.1 (Strategyproofness). A voting rule f is *strategyproof* if no single voter has a manipulation. That is, in every profile $\boldsymbol{L} \in \mathcal{L}(A)^n$, for every voter $i \in N$ and alternative vote $L'_i \in \mathcal{L}(A)$,

$$f(L'_i, \boldsymbol{L}_{-i}) \preceq_i f(\boldsymbol{L}).$$

We say that a voting rule is *manipulable* if it is not strategyproof. We can also rephrase this definition using game theory terminology:

Observation 3.2 A voting rule f is strategyproof if and only if for any preference profile $\boldsymbol{L}$ and any voter i, L_i is a weakly dominant strategy in the game $\langle f, \boldsymbol{L} \rangle$.

Proposition 3.3 *If there are exactly two candidates, then the Plurality voting rule is strategyproof.*

Proof. Denote $A = \{a, b\}$ and assume that candidate a was selected by the Plurality voting rule (for some given preference profile $\boldsymbol{L}$). This assumption is without loss of generality (henceforth, *w.l.o.g.*) since the proof if b is selected is symmetric.

Let $i \in N$ be one of the voters. If $b \prec_i a$, then voter i's favorite candidate has already won, and she certainly has nothing to gain by lying about her preferences. On the other hand, if $a \prec_i b$, then changing voter i's preference so that b is ranked below a only lowers b's score and cannot possibly cause b to win. In either case, voter i cannot benefit from lying about her preferences. □

Note that most common voting rules coincide with the Plurality rule when there are two candidates, so the proposition extends to all of them as well. We next define some useful properties of voting rules and prove several general lemmas.

Definition 3.4 (Maskin monotonicity). An SCC F is *Maskin monotone*[1] if for all $\boldsymbol{L}$, all $a \in F(\boldsymbol{L})$, all $i \in N$, and all L'_i s.t. $a \succ_i b \Rightarrow a \succ'_i b$ (that is, voter i only moves a up her ranking, possibly changing order between other candidates), it holds that $a \in F(\boldsymbol{L}_{-i}, L'_i)$.

The first lemma says that a strategyproof voting rule's selected outcome remains constant for all changes to the preference profile such that candidates ranked below the winner before the change are also ranked below the winner after the change:

Lemma 3.5 *Let f be a strategyproof voting rule, then f is Maskin monotone.*

Proof. See Exercise 3.6(1). □

Consider again the case of two alternatives. Plurality (sometimes referred to as *Majority* in this case) is not the only strategyproof rule. The only property of Majority used in the proof of Prop. 3.3 is its monotonicity—the fact that ranking a candidate higher never causes this candidate to lose. The proof thus shows that *any* monotone binary function is strategyproof. That is, the converse of Lemma 3.5 holds when there are only two candidates. For example, we

[1]There is also a weaker notion of monotonicity in the literature, where voters move up a without changing the order among other candidates. Most common voting rules are monotone under the latter definition (STV is a notable exception) but are not Maskin monotone. For $m = 2$, however, the two definitions coincide.

can think of a rule where a wins if it gets at least $n/3$ votes, or a rule where b wins unless voters 6 and 11 vote for a.

The second lemma says that the outcome of a Maskin monotone voting rule (and thus any strategyproof voting rule as well) must be Pareto optimal, meaning there is no candidate strictly preferred by all voters to the winning candidate:

Lemma 3.6 *Let f be a Maskin monotone voting rule which is onto, then f is Pareto.*

Proof. Assume, toward a contradiction, that there is a preference profile $\boldsymbol{L}$, candidates $a, b \in A$ s.t. $\forall i \in N : b \prec_i a$, and yet $f(\boldsymbol{L}) = b$. Since f is onto, there exists a preference profile $\boldsymbol{L}'$ such that $f(\boldsymbol{L}') = a$. Let $\boldsymbol{L}''$ be a preference profile where all voters rank candidate a first and candidate b second. We illustrate the three preference profiles below.

$\boldsymbol{L}$	$\boldsymbol{L}''$	$\boldsymbol{L}'$
$\cdots a \succ_1 \cdots \succ_1 b \cdots$	$a \succ''_1 b \cdots$	$\cdots a \cdots$
$\cdots a \succ_2 \cdots \succ_2 b \cdots$	$a \succ''_2 b \cdots$	$\cdots a \cdots$
$\cdots a \succ_3 \cdots \succ_3 b \cdots$	$a \succ''_3 b \cdots$	$\cdots a \cdots$
$\cdots$	$\cdots$	$\cdots$
$f(\boldsymbol{L}) = b$	$f(\boldsymbol{L}'') = ?$	$f(\boldsymbol{L}') = a$

No voters ranked b lower in $\boldsymbol{L}''$ than they did in $\boldsymbol{L}$ (based on the assumption that all voters ranked a above b in $\boldsymbol{L}$), so by Maskin monotonicity it follows that $f(\boldsymbol{L}'') = f(\boldsymbol{L}) = b$. On the other hand, no voters ranked a lower in $\boldsymbol{L}''$ than they did in $\boldsymbol{L}'$ (since a is always ranked first in $\boldsymbol{L}''$), so by Maskin monotonicity it follows that $f(\boldsymbol{L}'') = f(\boldsymbol{L}') = a$, which is a contradiction (since $a \neq b$ and f is a function). Hence, $f(\boldsymbol{L}) \neq b$. □

Lemma 3.7 *Let f be a voting rule which is Pareto, then f is unanimous.*

Proof. Consider the profile $\boldsymbol{L} = (L, L, \ldots, L)$ for some $L \in \mathcal{L}(A)$. The candidate $\text{top}(L)$ strictly Pareto-dominates any other candidate in A. Thus by Pareto, $f(\boldsymbol{L}) = \text{top}(L)$. □

3.2 THE GIBBARD–SATTERTHWAITE THEOREM

Definition 3.8 (Dictator). A voting rule f is *dictatorial-in-range* if there is an individual (the dictator) and a subset $B_i \subseteq A$ such that i's most preferred candidate in B_i is always chosen:

$$\exists i \in N \quad \forall \boldsymbol{L} \in \mathcal{L}(A)^n : f(\boldsymbol{L}) = \text{top}(L_i|_{B_i}).$$

Note that if there is such a dictator i, then range$(f) = B_i$. A voting rule f is called *dictatorial* if it is both dictatorial-in-range and onto (and thus $B_i = A$).

Definition 3.9 (Duple). A voting rule f is a *duple* if there are only two possible winners, that is, $|\text{range}(f)| \leq 2$.

Both dictatorial rules and duples have significant shortcomings as voting rules. A dictatorship ignores the will of all voters but one, and a duple may fail to select a candidate even if there is a unanimous agreement among voters that it is best. Unfortunately, if we require strategyproofness then no other rules are possible.

Theorem 3.10 (The G-S theorem [Gibbard, 1973, Satterthwaite, 1975]). *A deterministic and onto voting rule for at least three candidates is strategyproof if and only if it is dictatorial.*

It is easy to see that a dictatorial rule is strategyproof, since the dictator is always best off reporting the truth and all other voters have no power; if we allow duples, we can arbitrarily select two candidates a, b and hold a majority vote between them. First, we will prove two useful lemmas. We can rephrase the "difficult side" of the G-S theorem as follows.

Corollary 3.11 *If a deterministic voting rule is strategyproof, then it is either dictatorial-in-range or a duple.*

Proof. A voting rule that is not a duple must be onto on some range $B \subsetneq A$ of size $|B| \geq 3$. Also, if f is strategyproof it must ignore any candidate in $A \setminus B$: if there are two profiles $\boldsymbol{L}, \boldsymbol{L}'$ that only defer in how i ranks candidates from $A \setminus B$, but $f(\boldsymbol{L}) \neq f(\boldsymbol{L}')$, then either i has a manipulation from L_i to L'_i or vise versa.

Thus, f is equivalent to some onto voting rule $f_B : \mathcal{L}(B)^n \rightarrow B$. By the G-S theorem f_B must have some dictator $i \in N$, and thus f is dictatorial-in-range for the same dictator i and the range $B_i = B$. □

There are many different proofs of the Gibbard–Satterthwaite theorem; each one highlights some other property of strategyproof rules, or important connections with another topic. Several such proofs are surveyed in Barberà [2011] along with an extensive historical background.

We lay out a proof based on the monotonicity property. The result that every Maskin monotone rule must be dictatorial is known as the *Muller-Satterthwaite theorem* [Muller and Satterthwaite, 1977]. Svensson [1999] provided a simple proof for $n = 2$ voters that is relatively easy to understand without background. We follow the proof scheme of Svensson for the G-S theorem that relies on the Muller-Satterthwaite theorem for $n = 2$ (Prop. 3.12) and Lemmas 3.5

and 3.6. In other places in the book we cite theorems that generalize the G-S theorem (in one of its two forms here), thereby providing additional proofs of the theorem.

Proposition 3.12 **[Muller and Satterthwaite, 1977, Svensson, 1999]** *Let f be a deterministic voting rule for $n = 2$ and at least 3 candidates, that is both Maskin monotone and onto, then f is dictatorial.*

Proof. Let $\boldsymbol{L}$ be a preference profile and let $a, b \in A$ such that:

$$\forall x \in A \setminus \{a, b\} : (a \succ_1 b \succ_1 x) \wedge (b \succ_2 a \succ_2 x).$$

By Pareto optimality, $f(\boldsymbol{L}) \in \{a, b\}$. In the first case we consider, $f(\boldsymbol{L}) = a$. Now consider a preference L'_2 which satisfies:

$$\forall x \in A \setminus \{a, b\} : b \succ'_2 x \succ'_2 a.$$

Due to Pareto optimality, $f(L_1, L'_2) \in \{a, b\}$. Due to strategyproofness, $f(L_1, L'_2) \preceq_2 f(\boldsymbol{L}) = a$, which entails $f(L_1, L'_2) = a$. Maskin monotonicity now implies that f will select a as the winner for any preference profile where voter 1 ranks a first. So voter 1 is a dictator for candidate a. In the second case where $f(\boldsymbol{L}) = b$, we would get that voter 2 is a dictator for b. W.l.o.g. we assume that the first case holds for f. The analysis above can be repeated for all pairs of candidates $x, y \in A$, to show that:

$$\forall x, y \in A, \text{ either voter 1 is a dictator for candidate } x \text{ or voter 2 is a dictator for candidate } y. \qquad (\#)$$

We denote by $\boldsymbol{L}_{xy}$ an arbitrary profile where $\text{top}(L_1) = x, \text{top}(L_2) = y$.

For $i \in \{1, 2\}$, let $D_i \subseteq A$ denote the set of candidates for whom voter i is a dictator. To complete the proof, we must show that $D_1 = A$. That is, that voter 1 is a dictator for all candidates.

We first show that $|D_1 \cup D_2| \geq 2$. This follows immediately from (#) and the fact that $m \geq 3$.

Since $a \in D_1$, we get that D_2 is empty: otherwise, there are some $x \neq y$ such that $x \in D_1$ and $y \in D_2$. In the profile $\boldsymbol{L}_{xy}$ both x and y must win, which is impossible.

Finally, assume toward a contradiction that $D_1 \subsetneq A$; then there is $z \in A \setminus D_1$. By statement (#) above applied to the pair z, a, either $z \in D_1$ or $a \in D_2$. However, we already concluded that D_2 is empty, so we reach a contradiction, and voter 1 is a dictator (if we had initially assumed that $f(\boldsymbol{L}) = b$, voter 2 would have been the dictator). □

Corollary 3.13 *Let f be a deterministic voting rule for $n = 2$ and at least 3 candidates, that is both strategyproof and onto, then f is dictatorial.*

Proof. Follows as an immediate corollary of Proposition 3.12 and Lemmas 3.5 and 3.6. □

We continue with the full proof (roughly following Svensson [1999]).[2]

Proof of Theorem 3.10. Assume by induction that the theorem holds for n voters or fewer (thus, Corollary 3.13 proves the base case), and consider any strategyproof and onto voting rule f for $n+1$ voters and $m \geq 3$ candidates.

Throughout the proof we define several auxiliary voting rules. We will use g for voting rules with two voters, and h for voting rules with n voters.

We define the following voting rule g for two voters:

$$g(L_1, L_2) \triangleq f(L_1, L_2, L_2, \ldots, L_2).$$

Step 1: g is dictatorial Note that by Lemma 3.6, f is also Pareto and thus unanimous by Lemma 3.7. Since f is unanimous, g is also unanimous and in particular onto. We argue that g is strategyproof. Otherwise, there is a profile (L_1, L_2) with a manipulation for one of the voters. Clearly a manipulation L'_1 for voter 1 is also a manipulation in f, which would lead to a contradiction. Thus suppose that there is a manipulation L'_2 s.t.

$$f(L_1, L'_2, \ldots, L'_2) = g(L_1, L'_2) \succ_2 g(L_1, L_2) = f(L_1, L_2, \ldots, L_2) = f(\boldsymbol{L}). \tag{3.1}$$

For any $k \in [0, n]$, let $\boldsymbol{L}^k$ be the profile where the first $n+1-k$ votes vote as in $\boldsymbol{L}$, and the other k voters vote L'_2. Due to strategyproofness of f,

$$f(\boldsymbol{L}^{k+1}) = f(L_1, L_2, \ldots, L_2, L'_2, L'_2, \ldots, L'_2) \preceq_2 f(L_1, L_2, \ldots, L_2, L_2, L'_2, \ldots, L'_2) = f(\boldsymbol{L}^k)$$

for all $k \in [0, n-1]$. In particular, we get that

$$f(\boldsymbol{L}) = f(\boldsymbol{L}^0) \preceq_2 f(\boldsymbol{L}^n) = f(L_1, L'_2, \ldots, L'_2),$$

in contradiction to Equation (3.1). Thus, g is strategyproof. Since it is also onto, then by Cor. 3.13 it is dictatorial.

Step 2: the case where voter 1 is the dictator If the dictator is voter 1, then voter 1 is a dictator of f whenever all other agents report the same preference L_2. Let $\boldsymbol{L} \in \mathcal{L}(A)^{n+1}$ be an arbitrary profile. Denote $a = \text{top}(L_1)$ and let $\overline{L}_2$ be some preference where a is ranked last. Let $\overline{\boldsymbol{L}}^k \in \mathcal{L}(A)^{n+1}$ be the profile where the last k voters vote $\overline{L}_2$, and all other voters vote as in $\boldsymbol{L}$. Since changing from $\overline{L}_2$ to L_i must not be a manipulation for a voter with preferences $\overline{L}_2$, we have:

$$a = f(L_1, \overline{L}_2, \ldots, \overline{L}_2) = f(\overline{\boldsymbol{L}}^n) \succeq_{\overline{L}_2} f(\overline{\boldsymbol{L}}^{n-1}) \succeq_{\overline{L}_2} \cdots \succeq_{\overline{L}_2} f(\overline{\boldsymbol{L}}^0) = f(\boldsymbol{L}).$$

However, since a is the least preferred candidate in $\overline{L}_2$, all weak relations above must be equalities, and thus $f(\boldsymbol{L}) = a = \text{top}(L_1)$, which means that voter 1 is a dictator for f.

[2]There were some gaps in the proof of Svensson [1999] to which I saw no trivial justification, and thus rewrote parts of the proof. I thank Omer Lev for pointing out some of the oversights and how to fix them.

Step 3: the case where voter 1 is not the dictator Thus suppose that voter 2 is a dictator of g. Let $L_1^* \in \mathcal{L}(A)$ be some fixed preference order, and consider the following voting rule for n voters:

$$h_{L_1^*}(L_2, \ldots, L_{n+1}) \triangleq f(L_1^*, L_2, \ldots, L_{n+1}).$$

Step 3.1: every rule h is dictatorial Since any manipulation in $h_{L_1^*}$ yields a manipulation by the same voter in f, the rule $h_{L_1^*}$ must be strategyproof. To see that $h_{L_1^*}$ is onto, consider some candidate $a \in A$, and let L_2' where a is ranked first. Then

$$h_{L_1^*}(L_2', \ldots, L_2') = f(L_1^*, L_2', \ldots, L_2') = g(L_1^*, L_2') = \text{top}(L_2') = a,$$

since voter 2 is a dictator of g. Since $h_{L_1^*}$ is strategyproof and onto, then by our induction hypothesis it is dictatorial. As the identity of the dictator may depend on L_1^*, we denote the dictator of $h_{L_1^*}$ by $i_{L_1^*} \in [2, n+1]$.

Reversing the definition of h, we can write for any preference profile $\boldsymbol{L} \in \mathcal{L}(A)^{n+1}$

$$f(\boldsymbol{L}) = f(L_1, \boldsymbol{L}_{-1}) = h_{L_1}(\boldsymbol{L}_{-1}) = \text{top}(L_{i_{L_1}}).$$

That is, voter 1 essentially "selects" a dictator i_{L_1}, which then determines the outcome.

Step 3.2: all rules h have the same dictator It is left to show that there is some $i^* \in [2, n+1]$ such that $i_{L_1} = i^*$ for all L_1, in which case voter 1 has no influence, and voter i^* is a dictator of f. Suppose otherwise, then there are preferences L_1, L_1' such that $i = i_{L_1}, i' = i_{L_1'}$ and $i \neq i'$. Consider a preference order $L_{i'}$ where $\text{top}(L_{i'}) = \text{top}(L_1) = a$, and another preference order L_i where $\text{top}(L_i) = b \neq a$. Let $\boldsymbol{L}_{-1,i,i'}$ be an arbitrary profile of all other $n-2$ voters. We have that

$$f(L_1, L_i, L_{i'}, \boldsymbol{L}_{-1,i,i'}) = \text{top}(L_{i_{L_1}}) = \text{top}(L_i) = b,$$

whereas

$$f(L_1', L_i, L_{i'}, \boldsymbol{L}_{-1,i,i'}) = \text{top}(L_{i_{L_1'}}) = \text{top}(L_{i'}) = a \succ_1 b,$$

which means that voter 1 has a manipulation in favor of a. This contradicts our assumption that f is strategyproof. Thus voter i^* is a dictator of f and we are done. □

In Chapter 4 we study some relaxations of the model that lead to positive results. In the remainder of this chapter we consider three extensions that show the robustness of the negative result: The first demonstrates lower bounds for the frequency of manipulation (Section 3.3), and the second provides upper bounds on the minimal size of a coalition that can manipulate the outcome (Section 3.4). The bounds are interrelated in the sense that most voting rules can be very often manipulated by a sufficiently large coalition. The third extension (Section 3.5) considers irresolute voting rules that may return multiple winners.

3.3 FREQUENCY OF MANIPULATION

While the G-S theorem shows that for every rule there is a manipulation in *some* profile, perhaps there are rules for which those manipulations are too rare for us to care about. If, for example, there are only one or two "dangerous" profiles out of $(m!)^n$ then maybe we should be content with such "almost-strategyproof" rules.

We will show that in every reasonable voting rule there will be many more manipulations than that, but first we need a way to measure how frequent manipulations are.

The most common distribution of profiles that is studied in the literature is called *the impartial culture*, in which each of the $(m!)^n$ profiles is sampled with the same probability. For other distributions that aim to capture more realistically the preferences in a diverse population, see Berg [1985].

We define the *manipulation power* of voter i in rule f as

$$M_i(f) \triangleq \frac{1}{(m!)^n} \sum_{\boldsymbol{L} \in \mathcal{L}(A)^n} [\![\exists L'_i \text{ s.t. } f(\boldsymbol{L}_{-i}, L'_i) \succ_i f(\boldsymbol{L})]\!],$$

that is, as the probability that in a random profile (sampled uniformly according to the impartial culture) voter i will have some beneficial manipulation.

Note that we can rephrase the negative part of the G-S theorem (see Cor. 3.11) as follows.

Theorem 3.14 *For deterministic voting rule f, either f is dictatorial-in-range, or a duple, or* $\sum_{i \in N} M_i(f) > 0$.

Now that we have a quantitative measure for manipulation, we would like to similarly relax the notions of dictatorships and duples. We first define a distance measure between voting rules.

Definition 3.15 (Closeness). The *closeness* of voting rules f, g is defined as $C(f,g) = \frac{1}{(m!)^n} \sum_{\boldsymbol{L} \in \mathcal{L}(A)^n} [\![f(\boldsymbol{L}) = g(\boldsymbol{L})]\!]$, that is, the fraction of input profiles on which they agree.

Definition 3.16 (ε-bad). We say that a rule f is ε-bad if there is some other voting rule g_f s.t.

1. g_f is either a duple or dictatorial-in-range and
2. $C(g_f, f) \geq 1 - \varepsilon$.

In particular, dictatorial functions and duples are 0-bad. The question of whether we can lower bound the manipulation power of ε-bad rules for $\varepsilon > 0$ is a relatively recent one. The first studies of similar questions [Conitzer and Sandholm, 2006, Procaccia and Rosenschein, 2007] were more algorithmic in nature, and provided algorithms that with high probability can tell whether a manipulation exists or not. Xia and Conitzer [2008] showed bounds on the manipulation power of coalitions (see more details below).

Friedgut et al. [2008] were the first to prove such a lower bound on the manipulation power of a single manipulator (under a slightly different definition), albeit only for Neutral rules with $m = 3$. The state-of-the-art result now shows that:

Theorem 3.17 **[Mossel and Rácz, 2012].** *For any $\varepsilon \geq 0$, and any deterministic voting rule f, either f is ε-bad, or $\sum_{i \in N} M_i(f) > p(\frac{1}{n}, \frac{1}{m}, \varepsilon)$ where p is some polynomial function with positive coefficients.*

That is, to keep a low probability of manipulation, we must select a rule that is very close to being a duple or dictatorial. Note, for example, that all anonymous rules that respect voters' majority (including all scoring rules, Copeland, STV, etc.) are very far from being "bad" and thus have at least a polynomial probability of manipulation.

Also note that we get the difficult part of G-S theorem (Theorem 3.14) as a special case for $\varepsilon = 0$.

3.4 GROUP MANIPULATIONS

One may claim that Theorem 3.17 is insufficient to argue that manipulations are common. Indeed, a polynomial fraction may still be quite small. Even in the Plurality rule, which is the most manipulable, the probability of manipulation is about $\frac{1}{\sqrt{n}}$. When there are millions of voters, this becomes negligible.

However, things look less rosy when we consider groups of voters that can manipulate together. If we allow sufficiently large groups, then we will almost always have some groups (or even many groups) who can get a better outcome for all voters in the group by manipulating together.

Definition 3.18 **(Group manipulation).** Voting rule f has a *group manipulation of size* k under profile $\boldsymbol{L}$ if there is a set of voters $I \subseteq N$ of size $|I| \leq k$, and $\boldsymbol{L}'_I$ s.t. for all $i \in I$, $f(\boldsymbol{L}_{-I}, \boldsymbol{L}_I) \succ_i f(\boldsymbol{L})$. f is *group strategyproof* if there is no group manipulation of any size under any profile.

For example, consider a profile $\boldsymbol{L}$ with four candidates $A = \{a, b, c, d\}$. Denote by N_{xy} voters who rank x first and y second. In the profile $\boldsymbol{L}$ we have $|N_{ab}| = 100$, $|N_{ba}| = 80$, $|N_{ca}| = |N_{cb}| = 15$ and $|N_{da}| = |N_{db}| = 10$. So if the voting rule is Plurality, a wins and no single voter can change the outcome. However the set $I = N_{cb} \cup N_{db}$ has a group manipulation that ranks b first.

There is a tight connection between the average-case manipulability and the minimal size of the coalition that is required to change the outcome. For a wide class of voting rules known as "generalized scoring rules," and which contains most common voting rules, Xia and Conitzer [2008] showed the following: (i) "large coalitions" with substantially more than $\sqrt{n}$ voters (formally, of size at least $k = \Omega(n^\alpha)$ for $\alpha > \frac{1}{2}$) can decide the identity of the winner in almost every profile; (ii) for smaller coalitions (formally, of size $k = O(n^\alpha)$ where $\alpha < \frac{1}{2}$), the probability they have any effect on the outcome is small.[3]

The critical regime where the size of the manipulating coalition is $k = c\sqrt{n}$ for some constant c was studied by Mossel et al. [2013], which also give a nice review on the topic. Focusing again on generalized scoring rules, they showed that the probability that a group manipulation of size k exists grows smoothly with c from 0 to 1.

Mossel et al. also extend result (i) from [Xia and Conitzer, 2008] on large coalitions to any anonymous voting rule (under mild restrictions).

3.4.1 SAFE MANIPULATIONS

A group manipulation requires some level of coordination and trust that may not be present among voters with conflicting interests. A particularly special case where coordination may be easier is when all the manipulators have exactly the same preferences and plan to cast the same vote. For example, a supporter of a certain party may post a wide call to voters with similar interest to cast some strategic vote in hope that enough voters will follow to change the outcome.

However, posting such a call bears a risk. Suppose the manipulator calculated that (say) 1,000 followers will change the outcome in some beneficial way. If too few people or too many people follow the suggested vote, it might have an adverse affect on the outcome that will hurt the manipulators.

Example 3.19 [Slinko and White, 2014] Suppose 41 voters are using the Borda rule to select one of five alternatives $A = \{a, b, c, d, e\}$. Let sincere preferences be distributed as follows.

Preference order	$abcde$	$cebad$	$ebcad$	$edacb$
Number of voters	10	15	14	2

When all voters state their sincere preferences, a scores 73, b 102, c 110, d 16, and e 109; c wins. Consider a voter i with $L_i = abcde$ (a shorthand for $a \succ_i b \succ_i c \succ_i d \succ_i e$). Alone, i cannot affect the outcome and gain. Yet, if at least 8 voters with same preferences as i vote $L'_i = badce$ then, ceteris paribus, b wins. Since $b \succ_i c$, voter i may post a call to convince other voters of the same type to vote L'_i.

However, if between two and six voters of type L_i vote L'_i, ceteris paribus, then e wins. As $e \prec_i c$, the manipulation hurts voter i.

[3] By small here we mean "goes to 0 as n grows," without specifying the rate of convergence which may be polynomial. However, in the part above we considered such probabilities (polynomial rather than exponential) as large. All is relative in life.

Slinko and White provide additional examples where *too many* followers result in an inferior outcome.

Definition 3.20 (Safe manipulation). A *safe manipulation* in profile $\boldsymbol{L}$, is a vote L'_i of voter i, such that:

1. for some subset I of voters of type L_i, $f(\boldsymbol{L}_{-I}, \boldsymbol{L}'_I) \succ f(\boldsymbol{L})$; and
2. for any subset I of voters of type L_i, $f(\boldsymbol{L}_{-I}, \boldsymbol{L}'_I) \succeq f(\boldsymbol{L})$, where $\boldsymbol{L}'_I = (L'_i, \ldots, L'_i)$.

That is, it is possible to gain if the "right" set of followers casts L'_i, but no set of followers may cause harm.

The main result of Slinko and White [2014] is an extension of the G-S theorem to safe manipulations.

Theorem 3.21 [Slinko and White, 2014]. *A deterministic and onto voting rule for at least three candidates has no safe manipulations if and only if it is dictatorial.*

As with the G-S theorem (but with some more effort) we can extend Theorem 3.21 as follows.

Corollary 3.22 *If a deterministic voting rule has no safe manipulations, then it is either dictatorial-in-range or a duple.*

We first prove the following lemma.

Lemma 3.23 *Let f be a deterministic voting rule with no safe manipulations. If $x \in A$ is not in the range of f, then f ignores the position of x (i.e., for any $\boldsymbol{L}, \boldsymbol{L}'$ s.t. $a \succ_i b \iff a \succ'_i b$ for all $i \in N$, $a, b \in A \setminus \{x\}$, it holds that $f(\boldsymbol{L}) = f(\boldsymbol{L}')$).*

Proof. Assume toward a contradiction that there are two profiles $\boldsymbol{L}, \boldsymbol{L}'$ that differ only in the position of x, s.t. $f(\boldsymbol{L}) = a$, $f(\boldsymbol{L}') = b$ and a, b, x are distinct candidates (this is w.l.o.g. as x is not in the range). W.l.o.g. we can also assume that $\boldsymbol{L}, \boldsymbol{L}'$ differ by a single voter i, as we can switch voters one by one. Clearly, if $b \succ_i a$ then L'_i is a manipulation at $\boldsymbol{L}$, and otherwise L_i is a manipulation at $\boldsymbol{L}'$. We will assume $a \succ_i b$ (the proof in the other case is symmetric).

Let k be the smallest number such that the scenario above occurs, and there are k voters with preference L_i. We will argue that i has a safe manipulation (which is a contradiction). Denote the set of voters with preference L_i by K.

First, if $k = 1$ then i is the only voter with preference L_i, so the manipulation L'_i is safe. If $k > 1$, we define for all $I \subseteq K \setminus \{i\}$ the profile $\boldsymbol{L}^I$ to be the profile $\boldsymbol{L}'$ where all voters in J switched to L'_i. In particular, $\boldsymbol{L}^{\emptyset} = \boldsymbol{L}'$. Recall that $f(\boldsymbol{L}^{\emptyset}) = b$. For any nonempty set $I \subseteq K$, consider some $j \in I$, and the profiles $\boldsymbol{L}^I, \boldsymbol{L}^{I \setminus \{j\}}$. These profiles differ by the position of x in a single voter's preference. Thus by our minimality assumption, they have the same outcome. Applied to all subsets I, we get that $f(\boldsymbol{L}^I) = f(\boldsymbol{L}^\emptyset) = b \succ_i a = f(\boldsymbol{L})$ for all $I \subseteq K$. Thus i has a safe manipulation, in contradiction to our initial assumption. □

***Proof of Cor.* 3.22.** The only case not already covered by Theorem 3.21 is a deterministic voting rule that has no safe manipulations, is non-dictatorial, not a duple, and not onto. We will show that such a voting rule cannot exist. Assume toward a contradiction that such a voting rule $f : \mathcal{L}(A)^n \to A$ exists. Since f is neither a duple nor onto, it has a range $B \subsetneq A$ of size $|B| \geq 2$. Since f has no safe manipulations, by Lemma 3.23 it must ignore the position of candidates in $A \setminus B$. Thus, there is a voting rule $f_B : \mathcal{L}(B)^n \to B$ such that for any $\boldsymbol{L} \in \mathcal{L}(A)$, $f_B(\boldsymbol{L}|_B) = f(\boldsymbol{L})$, where $\boldsymbol{L}|_B$ is the profile $\boldsymbol{L}$ with candidates from $A \setminus B$ removed.

In particular, f_B is a deterministic onto voting rule for $|B| \geq 3$ candidates with no safe manipulations, and thus by Theorem 3.21 must be dictatorial with some dictator i. Hence, f must be dictatorial-in-range for i and the range $B_i = B$. □

Safe manipulations make the assumption that voters only deliberate among two possible actions: being truthful or playing some fixed manipulation L_i'. Elkind et al. [2015a] make a similar assumption and consider the $n \times 2$ game induced by these actions.

3.5 IRRESOLUTE SOCIAL CHOICE CORRESPONDENCES

Consider an SCC that returns a set of winners rather than a single winner, for example, F^{TOPS} which returns the set of all candidates ranked first by some voter. We may ask if a voter has a manipulation under F^{TOPS}.

A key observation is that an ordinal preference over alternatives does not induce a complete preference order over *sets of candidates*. For example, suppose that $a \succ_i b \succ_i c$ according to $L_i \in \mathcal{L}(A)$. Does voter i prefer alternative b over the set $\{a, c\}$? We deal with a similar problem when considering *randomized outcomes* later in Section 4.3.

In order to have a well-defined notion of strategyproofness for irresolute SCCs, we need some way to extend preferences over alternatives to preferences over sets. We can also settle for *partial preferences* (or any other binary relation), with the implicit assumption that a voter will not use a manipulation L_i' if $F(\boldsymbol{L}_{-i}, L_i')$ and $F(\boldsymbol{L})$ are incomparable.

One way of obtaining (partial) preference orders over sets that is common in the decision-making literature is to impose various axioms (see Cho [2012], Geist and Endriss [2011]). In the context of SCCs, we can sometimes think of these axioms as beliefs that the voter has on the tie-breaking procedure [Meir, 2016c].

For example, the natural requirement that if all options in B are strictly preferred to all options in B' then $B \succ B'$ is known as the (strict) *Kelly axiom* [Kelly, 1977]. Suppose that we extend the preferences of all voters to sets using *only* the strict Kelly axiom, leaving all other preferences incomparable (this corresponds to a voter who believes ties will always be broken against her when manipulating, but in her favor when truthful).

Observation 3.24 The Plurality SCC (denoted by F^P) is strategyproof under the strict Kelly preference extension.

Proof. Suppose $W = F^P(\boldsymbol{L})$. Let $i \in N$, $L'_i \neq L_i$, and $W' = F^P(\boldsymbol{L}_{-i}, L'_i)$. If $\text{top}(L_i) \in W$ then no W' can be better for i. Otherwise, $W' \subseteq W$. This means that $W \cap W' \neq \emptyset$ and thus W, W' are either the same or incomparable. In either case, W' is not preferred over W.

□

Another approach that is more standard in the economic literature is to impose *cardinal utilities*, and then assume some randomized tie-breaking method (e.g., select uniformly from the set of winners). To formally define a notion of manipulation in (ordinal) SCCs, we then need to decide two things: first, if the voter with preferences L_i has to gain for *every* U_i that fits L_i, or just for some cardinal extensions; second, does the voter have to gain with respect to all tie-breaking lotteries, some lotteries, and so on. Any such combination induces a different notion of manipulation, and thus a different characterization of strategyproof SCCs.

The Duggan–Schwartz theorem We provide a strong negative result for one such preference extension, taken from Duggan and Schwartz [2000]. A *tie-breaking dominance (TBD) manipulation* allows the manipulator to gain in expectation according to a *single* utility function, but for *every* randomized tie breaking.

Formally, we say that a voter i with preferences L_i *TBD-prefers* the set W to W' if for any lotteries $p \in \Delta(W), p' \in \Delta(W')$ with full support, there is some U_i that fits L_i s.t. $E_{c \sim p}[U_i(c)] > E_{c \sim p'}[U_i(c)]$.

Definition 3.25 (TBD-strategyproofness). An SCC F is *TBD-strategyproof* if there is no profile $\boldsymbol{L}$, $i \in N$, and L'_i such that i TBD-prefers $F(\boldsymbol{L}_{-i}, L'_i)$ to $F(\boldsymbol{L})$.

For example, if $L_i = abcd$ then $W = \{a, b\}$ is TBD-preferred to $W' = \{b, c, d\}$, since $p(a) > 0$ and we can always set $U_i(a)$ high enough. In contrast, $W = \{a, b, d\}$ is not TBD-preferred to $W' = \{a, c, d\}$, e.g., if we set $p(d) = p'(a) = 0.9$: if we try to increase $U_i(b)$ to make W more attractive, we must also increase $U_i(a)$. We should note that TBD-preference is *not* a partial order, since it is not antisymmetric. For example, the sets $W = \{a, c\}$ and $W' = \{b\}$ are TBD-preferred to one another.

Residual resoluteness is a technical condition that requires $F(\boldsymbol{L})$ to be a singleton in profiles where almost all voters are unanimous. An irresolute SCC F is *strongly dictatorial* if there is a voter $i \in N$ such that $F(\boldsymbol{L}) = \{\text{top}(L_i)\}$ for all $\boldsymbol{L}$.

Theorem 3.26 [Duggan and Schwartz, 2000]. *Every TBD-strategyproof onto SCC F that satisfies residual resoluteness is strongly dictatorial.*

Note that when F is resolute, TBD-strategyproofness boils down to strategyproofness, and thus Theorem 3.26 implies the G-S theorem. An alternative formulation of the Duggan-Schwartz theorem which uses axiomatic preference extensions rather than lotteries is presented in Taylor [2002]: any onto SCC is either *weakly dictatorial* (the top preference of the dictator is always *one of* the winners), or there is an *optimistic manipulation* (where some $c \in F(\boldsymbol{L}, L'_i)$ is preferred to all of $F(\boldsymbol{L})$), or there is a *pessimistic manipulation* (where some $c \in F(\boldsymbol{L})$ is worse than all of $F(\boldsymbol{L}_{-i}, L'_i)$). Taylor also provides an extensive discussion and many other results on irresolute voting rules.

3.6 EXERCISES

1. Show that every strategyproof voting rule is Maskin monotone.
2. Give an example of a manipulation in the STV voting rule with 3 candidates. The correctness of the example should not depend on tie-breaking.
3. Consider 3 candidates and 600 voters in Plurality, whose preferences are sampled independent and identically distributed from the uniform distribution on $\mathcal{L}(A)$. Suppose that candidate a wins. Show that with probability of at least 95% there is a coalition of size (at most) 70 that has a manipulation for candidate b.
4. Compute the closeness of Plurality and Borda (i.e., $C(f^{PL}, f^{BL})$) for $m = 3$ and $n = 2$.
5. Give examples of a safe and an unsafe manipulation in Plurality-with-Runoff with 3 candidates, and at least 3 voters of each existing type.
6. Prove that every manipulation in Plurality is safe.
7. Give an example of a TBD-manipulation in the Plurality SCC F^P, that is not a manipulation for a voter with arbitrary fixed cardinal preferences.

CHAPTER 4

Regaining Truthfulness in Voting

It is important to understand what parts of the voting model are responsible for the strong negative results presented in the previous chapter.

We will focus on four approaches, each of which attains truthfulness by relaxing some assumption underlying the G-S theorem. Other approaches that involve monetary payments are discussed in Chapter 5. Yet another approach relaxes the assumption that voters know exactly how others will vote. We touched upon this in Section 3.4.1 when discussing safe manipulations, and we return to it in Section 8.1 when discussing uncertainty.

Section	*Approach*	*Relaxed Assumptions*
4.1	Domain restriction	Any strict preference profile $\boldsymbol{L}$ is possible
4.2	Complexity barriers	Unbounded computational resources
4.3	Randomized voting rules	Voting rule is deterministic
4.4	Almost-strategyproof rules	Voters will manipulate even for a small gain

4.1 DOMAIN RESTRICTION

The G-S theorem requires that all profiles in $\mathcal{L}(A)^n$ are allowed. We observe that some richness in the preferences is required for the negative result: suppose that $A = \{a, b, c\}$ and we know that no voter ever ranks b last. We can define a voting rule f^b that always selects b as a winner, unless there is a unanimous vote for a or for c. Clearly, if $f^b(\boldsymbol{L}) \in \{a, c\}$ then there is no manipulation. If $f^b(\boldsymbol{L}) = b$ and $f^b(\boldsymbol{L}_{-i}, L'_i) = a$ for some $i \in N$, then either $L_i = c \succ b \succ a$ or $\text{top}(L_i) = b$. In either case, $a \prec_i b$ so i does not gain. We get that f^b is strategyproof and onto.

We thus ask what other restrictions we can impose on the set of preferences to obtain strategyproof rules.

4.1.1 SINGLE-PEAKED PREFERENCES ON A LINE

Suppose voters are voting on where to place a public library along a street. Naturally, each voter prefers the library to be located as close as possible to her house (whose location is private information). If we number all the addresses in the street from (say) east to west, then it is not possible for example that a voter prefers the east-most location, then the west-most location, and then some location in the middle. Thus not every preference profile is possible.

Definition 4.1 (Single-peaked preferences on a line). Given an order $\mathcal{O}$ on candidates A, a (weak) preference order R_i is *single-peaked* with respect to $\mathcal{O}$ if there is some "peak candidate" a_i^* s.t. if x is *between* a_i^* and y then i prefers x over y. That is, for all $x, y \in A$, if $\mathcal{O}(a_i^*) \geq \mathcal{O}(x) > \mathcal{O}(y)$ or $\mathcal{O}(a_i^*) \leq \mathcal{O}(x) < \mathcal{O}(y)$, then $a_i^* \succsim_i x \succ_i y$. In particular, a_i^* is among i's most preferred candidates.[1]

A profile $\boldsymbol{R}$ is *single-peaked* if there is some order $\mathcal{O}$ s.t. every R_i is single-peaked w.r.t. $\mathcal{O}$.

See Figure 4.1 for an example.

The Median Voter rule Given a linear order $\mathcal{O}$ over alternatives A, we define the *Median voter rule* $f_{\mathcal{O}}$ as follows. For any preference profile $\boldsymbol{L}$ that is single-peaked on $\mathcal{O}$:

1. Ask each voter $i \in N$ to report her peak $a_i^* = \text{top}(L_i)$.
2. Set $l_i = \mathcal{O}(a_i^*)$, that is, the order of a_i^* in $\mathcal{O}$.
3. Find $j^* \in N$ s.t. $l_{j^*} \in \text{Median}\{l_1, \ldots, l_n\}$ (if more than one exists, break ties toward the left-most voter).
4. Return $f_{\mathcal{O}}(\boldsymbol{L}) = a_{j^*}^*$.

There are various ways to extend the mechanism to weak preferences profiles, which we do not consider here. Consider the example in Figure 4.1. The median of the five numbers $\{l_1, l_2, l_3, l_4, l_5\} = \{1, 2, 4, 5, 4\}$ is 4. Thus either voter 3 or voter 5 can be the median voter. In either case, the outcome is $\text{top}(L_3) = \text{top}(L_5) = D$, which is the *median candidate*.

L_1	$A \succ E \succ C \succ D \succ B$
L_2	$E \succ A \succ C \succ D \succ B$
L_3	$D \succ B \succ C \succ E \succ A$
L_4	$B \succ D \succ C \succ E \succ A$
L_5	$D \succ C \succ E \succ B \succ A$
L_6	$D \succ C \succ B \succ A \succ E$

L_1	L_2		L_5 L_3	L_4
A	E	C	D	B
1	2	3	4	5

Figure 4.1: The preferences of the first five voters are single-peaked w.r.t. the order $\mathcal{O} = A \succ E \succ C \succ D \succ B$. The right figure shows the position of each of the first five voters w.r.t. the order $\mathcal{O}$. For example, $l_4 = \mathcal{O}(B) = 5$. The sixth voter L_6 is not single-peaked w.r.t. $\mathcal{O}$.

[1] Note that this model is highly related to the facility location problem in Section 5.3. However, here we do not assume any metric distance or cardinal utilities.

Theorem 4.2 **[Black, 1948, Moulin, 1980].** *The Median Voter rule is group-strategyproof (in particular strategyproof).*

Proof. denote $f = f_{\mathcal{O}}$ and suppose that $\boldsymbol{L}$ is single-peaked w.r.t. $\mathcal{O}$. Assume toward a contradiction that there is a subset $I \subseteq N$ that can manipulate by reporting $\boldsymbol{L}'_I$, and denote $c^* = f(\boldsymbol{L}), c' = f(\boldsymbol{L}_{-I}, \boldsymbol{L}'_I)$. Clearly I does not contain any voter i s.t. $a^*_i = f(\boldsymbol{L})$ as such a voter strictly loses from any manipulation. Denote by $I_R, I_L \subseteq I$ the sets of agents whose peaks are to the right and to the left of c^*, respectively. W.l.o.g. $\mathcal{O}(c') < \mathcal{O}(c^*)$, that is, c' is to the left of c^*. It cannot be that I_R is empty, since no report by I_L can move the median left. However, consider some $i \in I_R$, then $\mathcal{O}(c') < \mathcal{O}(c^*) < \mathcal{O}(a^*_i)$ thus i strictly loses from this manipulation. This is a contradiction. □

It also holds that under single-peaked preferences the Condorcet winner always exists and is always the median candidate. Moulin goes on and characterizes all strategyproof anonymous voting rules under single-peaked preferences. Such a mechanism is always the median of the n voters and at most $n + 1$ other fixed points. If there are at most $n - 1$ fixed points then the mechanism is also onto. On the down side, even the Median rule does not satisfy the stronger incentive property of being *obviously stratregyproof* [Bade and Gonczarowski, 2017].

Note that if we relax anonymity there are many other strategyproof mechanisms. For example, we can divide the voters into several distinct groups $N_1, \ldots, N_k$, compute the median c^*_k of each group, and return the median of $c^*_1, \ldots, c^*_k$ ("median-of-medians").

4.1.2 OTHER SINGLE-PEAK DOMAINS

An important question in social choice is what other domain restrictions allow for strategyproof (and onto) rules. For example, the Median mechanism on a line can be extended to a *tree* and maintain its strategyproofness (see Exercise 4.5(2)).

Kalai and Muller [1977] provide an axiomatic characterization of all such restrictions, of which single-peakedness is a special case. However this characterization is not always helpful when we try to impose explicit restrictions or find whether a particular voting rule is strategyproof.

Nehring and Puppe [2007] take a more constructive approach (which builds on Barberà et al. [1997] and other previous work). They first generalize the notion of "betweenness" and thereby obtain a very broad class of "single-peaked spaces."

Intuitively, instead of sorting alternatives on a line, we can think of an arbitrary set of *properties* that alternatives may share, where each property is an arbitrary set $P \in 2^A \setminus \{A, \emptyset\}$. Then, alternative x is *between* y and z if it shares all the properties that both y and z share, that is, if $\{y, z\} \subseteq P \Rightarrow x \in P$ for every property P. Thus any set of properties $\boldsymbol{P} = \{P_1, P_2, \ldots\}$ induces a betweenness relation (this is a ternary relation, but we avoid a special notation for it).

Similarly, any betweenness relation that holds a certain set of axioms induces a set of properties $\boldsymbol{P}$.

Example 4.3 Consider five voters, Alice, Bob, Chen, David and Elena, who want to jointly rent a car. The available cars may have different binary properties, and in particular may be red/black (P_1); manual/automatic (P_2); and have two/four doors (P_3). We denote every binary property by 0/1. The peak preference of each voter is as follows:

	P_1	P_2	P_3
Alice	1	1	1
Bob	1	1	0
Chen	1	0	1
David	0	0	0
Elena	1	0	0

For example, Bob's favorite car is black, automatic, with two doors (110). All voters are trivially between Alice and David, since they share no properties. In addition, Elena is between Bob and Chen, since she shares the only property they share (black car). On the other hand, David is not between Bob and Chen since he prefers a red car.

Definition 4.1 extends naturally to any set A with a collection of properties $\boldsymbol{P} = \{P_1, P_2, \ldots\}$. To see why single-peakedness on a line is a special case, consider any order $\mathcal{O}$, and all properties of the form $P_{lx} = \{a : \mathcal{O}(a) \leq \mathcal{O}(x)\}$ or $P_{rx} = \{a : \mathcal{O}(a) \geq \mathcal{O}(x)\}$ for some $x \in A$. For example, for $x = E$ in Figure 4.1, we have the properties $P_{lE} = \{A, E\}$ ("being at or left to E") and $P_{rE} = \{E, C, D, B\}$ ("being at or right to E"). For another example see Exercise 4.5(3).

Without further requirements on the collection of properties $\boldsymbol{P}$, the betweenness relation and derived single-peakedness are not very interesting, as every preference order is single-peaked w.r.t. *some* $\boldsymbol{P}$.

Since the relations between candidates are determined in terms of properties, it makes sense to also define voting rules in term of properties. Associate with any property P a collection of non-empty subsets of voters $\mathcal{W}_P \subseteq 2^N \setminus \{\emptyset\}$.

Definition 4.4 (Voting by issues). Define $F_{\boldsymbol{P}} : \mathcal{L}(A)^n \to 2^A$ as follows: $x \in F_{\boldsymbol{P}}(\boldsymbol{L})$ if and only if $x \in P$ exactly for the properties P such that $\{i : \text{top}(L_i) \in P\} \in \mathcal{W}_P$.

This definition looks a bit cumbersome but has a very intuitive interpretation. Each property P has a "set of winning coalitions" $\mathcal{W}_P$ that can justify P. That is, the outcome $F_{\boldsymbol{P}}(\boldsymbol{L})$ must have property P if the set of voters who vote for (some alternative with) property P is a winning coalition for P, and otherwise the outcome should not have property P. Using the line as an example again, then the Median mechanism (with odd n) is a voting-by-issues (VBI) rule, where the properties are $\{P_{lx}, P_{rx}\}_{x \in A}$ and $\mathcal{W}_P$ contains all coalitions of size at least $\lceil \frac{n}{2} \rceil$. Note that if a majority of voters have their peak at or to the right of x, then indeed the median will also be at or to the right of x.

It is possible that $F_{\boldsymbol{P}}(\boldsymbol{L}) = \emptyset$. Consider Example 4.3, where $\mathcal{W}_{P_1} = \{N\}$, $\mathcal{W}_{P_2}$ contains all sets of at least three voters, and $\mathcal{W}_{P_3} = 2^N \setminus \{\emptyset\}$. Then only property P_3 (which is justified by any non-empty coalition) is justified in the profile $\boldsymbol{L}$. However, of the five available cars, there is no car that has *only* property P_3 (i.e., no car is red, manual, with four doors). Thus $F_{\boldsymbol{P}}(\boldsymbol{L}) = \emptyset$.

If non-empty, the outcome of $F_{\boldsymbol{P}}$ is always unique. If $F_{\boldsymbol{P}}$ is never empty it is called *consistent*, and thus consistent VBI rules are SCFs. Since $F_{\boldsymbol{P}}$ only makes use of the top preference (peak) of each voter, we can also think of a consistent VBI rule as a (resolute) function $F_{\boldsymbol{P}} : A^n \to A$.

Which restricted domains allow strategyproof voting? Given a restricted domain $\hat{\mathcal{L}} \subseteq \mathcal{L}(A)$, and a restricted voting rule $f : \hat{\mathcal{L}}^n \to A$, we may or may not be able to write f as a consistent VBI. The main result of Nehring and Puppe [2007] is an exact characterization of strategyproofness by this condition. A classification of such restricted domains is in Figure 4.2.

Theorem 4.5 [Nehring and Puppe, 2007]. *A restricted voting rule $f : \hat{\mathcal{L}}^n \to A$ is onto and strategyproof if and only if there are properties $\boldsymbol{P}$ and a VBI $F_{\boldsymbol{P}}$ such that: (i) $F_{\boldsymbol{P}}$ is consistent; and (ii) $F_{\boldsymbol{P}}(\boldsymbol{L}) = f(\boldsymbol{L})$ for all $\boldsymbol{L} \in \hat{\mathcal{L}}^n$.*

4.1.3 DICHOTOMOUS PREFERENCES

We say that the weak preferences $R_i \in \mathcal{R}(A)$ are *dichotomous* if we can partition A into two nonempty disjoint sets A_i^G, A_i^B such that voter i strictly prefers all alternatives in A_i^G ("Good") to those in A_i^B ("Bad") but is completely indifferent within each set. A preference profile $\boldsymbol{R}$ is dichotomous if all of R_i are. We denote by $\mathcal{D}(A) \subseteq \mathcal{R}(A)$ the set of dichotomous weak preferences. Note that each $R_i \in \mathcal{D}(A)$ can be written as a binary vector of length m, where "1" is a good alternative and "0" is a bad one.

One situation where dichotomous preferences naturally arise is *scheduling* (where each agent only cares whether the selected schedule meets her requirements). Another similar situation is choosing a software for shared use, which needs to be compatible with current software of each user. Dichotomous preferences have been studied in the context of the Approval voting system by Brams and Fishburn [1978].

A *non-ranked voting rule* f^B is defined by a set of allowed ballots $B \subseteq \{0, 1, \ldots, m-1\}$. Intuitively, each voter votes for an allowed number of candidates, that is, $a_i \in A_i \triangleq \{a \in 2^A : \|a\|_1 \in B\}$. The winner is the candidate with the most votes $f^B(\boldsymbol{a}) = \text{argmax}_{c \in A} \sum_{i \in N} [\![c \in a_i]\!]$, breaking ties lexicographically.[2] If $B = \{k\}$ is a singleton then f^B is the k-Approval rule, which is a valid voting rule (i.e., a standard game form). Otherwise, f^B is non-standard. The Approval voting rule is f^{AP} is the rule $f^{[m-1]}$, that is, where each voter

[2]In Brams and Fishburn [1978] there is no tie-breaking, and preferences are extended to sets of winners (see Section 4.3).

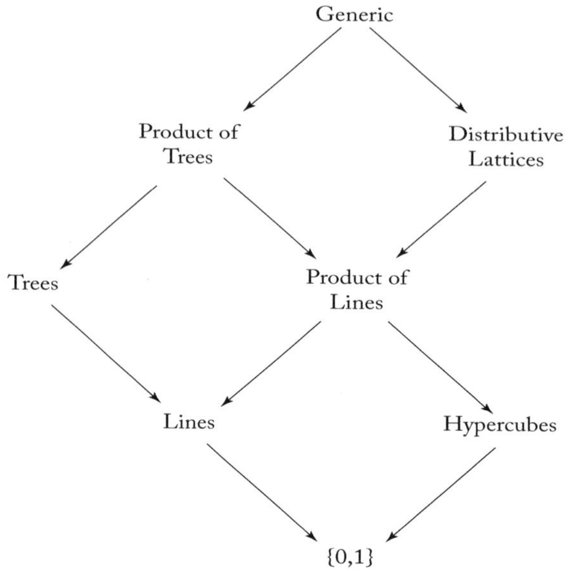

Figure 4.2: A diagram based on Nehring and Puppe [2007, p. 294] showing the relations between various classes of median spaces. An arrow from A to B means that B is a special case of A.

may vote for any subset of candidates. Note that for Approval, for any dichotomous preference $R_i \in \mathcal{D}(A)$ there is an action $a_i \in A_i$ that approves exactly the "good" candidates. We call this action the *truthful vote* of voter i. In general, a voter may not have a truthful action in a non-ranked rule f^B.

Brams and Fishburn focus on characterizing dominance relations between strategies. For dichotomous preferences, their characterization boils down to a simple result.

Theorem 4.6 [Brams and Fishburn, 1978]. *Under the Approval voting rule f^{AP}, it is a strictly dominant strategy for a voter with dichotomous preferences to vote truthfully.*

Proof. Let $\boldsymbol{R}$ be a weak preference profile where $R_i = (A_i^G, A_i^B)$ is dichotomous for some voter i. Denote $c = f^{AP}(\boldsymbol{R})$. If $c \in A_i^G$ then clearly i has no manipulation. Thus suppose that $c \in A_i^B$ and consider some alternative vote $R_i' = (A_i'^G, A_i'^B)$. In the profile $(\boldsymbol{R}_{-i}, R_i')$, candidates in $A_i'^G \setminus A_i^G \subseteq A_i^B$ get one more vote, and candidates in $A_i'^B \setminus A_i^B \subseteq A_i^G$ get one vote less. Thus $f^{AP}(\boldsymbol{R}_{-i}, R_i') \in A_i^B$, which means i does not gain. This settles that $f^{AP}(\boldsymbol{R}) \succsim_i f^{AP}(\boldsymbol{R}_{-i}, R_i')$ for any R_i'.

To show strict dominance, we need to show a profile where i strictly gains by playing R_i instead of R_i'. Suppose that there is $d \in A_i^G \setminus A_i'^G$ (otherwise we use a similar construction for

$d \in A_i^B \setminus A_i'^B$), and let $b \in A_i^B$. Consider a profile $\boldsymbol{R}_{-i}$ where both b, d have $n-1$ votes and all other candidates have 0. Otherwise, Consider a profile $\boldsymbol{R}_{-i}$ where d has $n-2$ votes and b has $n-1$ votes. If b has a tie-breaking advantage over d. In either case, $f^{AP}(\boldsymbol{R}_{-i}, R_i) = d \succ_i b = f^{AP}(\boldsymbol{R}_{-i}, R_i')$. □

The main result of Brams and Fishburn in the paper is that Approval always elects a weak Condorcet winner in dominant strategies under dichotomous preferences (in particular one exists), and that it is *the unique* non-ranked voting rule with this property. Vorsatz [2007] later showed that Approval is also the unique truthful voting rule under dichotomous preferences if we require neutrality, anonymity, and strict monotonicity. Compare with the characterization of Moulin for single-peaked domains from Section 4.1.1.

4.2 COMPLEXITY BARRIERS

Even though the G-S theorem states that manipulations exist under any voting rule, a voter trying to manipulate might find it difficult to know *how* to manipulate. This observation led to the idea that some voting rules might be truthful *in practice*, assuming that voters have limited computational resources. The link between manipulations and computational complexity was suggested and formalized by Bartholdi et al. [1989], who used the notion of *NP-hardness*.

We do not formally define here what is an NP-hard problem, and refer the reader to standard textbooks (e.g., Van Leeuwen [1991]) for definitions and further discussion. For our purpose, it will be enough to say that: (1) it is believed that there are no efficient algorithms (i.e., algorithms whose runtime is polynomial in the size of the input) for solving NP-hard problems; (2) to prove that a problem is NP-hard, one needs to show a *reduction* from another problem that is known to be NP-hard, such as the CLIQUE problem or the TRAVELING-SALESMAN problem [Karp, 1972].

Bartholdi et al. formalized the following computational problem, which can be applied to any voting rule f.

> MANIPULATION$_f$: given a set of candidates A, a group of voters N, a manipulator $i \in N$, a preference profile $\boldsymbol{L}_{-i}$ of all voters except i, and a specific candidate $p \in A$: Answer whether the manipulator can provide a preference L_i^* such that $f(\boldsymbol{L}_{-i}, L_i^*) = p$.

Then they asked whether there is a voting rule f such that computing the outcome $f(\boldsymbol{L})$ is easy but the problem MANIPULATION$_f$ is NP-hard. Note that since the number of possible reports is $m!$, a brute-force search is typically infeasible.

Notice the following observations.

- The problem is a decision problem. We would like to know if there *exists* such a preference.

- The problem is defined more strongly than the simple definition of manipulation. The question is not whether the manipulator can guarantee a victory of a more preferred candidate, but whether he can guarantee the victory of a *specific* candidate.

- The G-S theorem specifies that one of the voters can manipulate the results in some profile $\boldsymbol{L}$. We don't know if this is true under the given circumstances, where $\boldsymbol{L}$, i, and p were given.

At least for some voting rules, it is easy to tell whether a manipulation exists or not. For example, in the Plurality rule f^{PL} it is sufficient to let i rank p at the top of L_i^*, followed by all other candidates in an arbitrary order. A manipulation exists if and only if $f(\boldsymbol{L}_{-i}, L_i^*) = p$. Thus **MANIPULATION**$_{f^{PL}}$ can be solved in polynomial time. Similar greedy algorithms work for many other voting rules.

Voting rules that are NP-hard to manipulate Note first that in some rules, such as *Kemeny rule* [Kemeny, 1959], it is already NP-hard to compute the outcome of a given profile [Dwork et al., 2001]. Such rules are clearly hard to manipulate, but this is not an interesting result.

Theorem 4.7 [Bartholdi et al., 1989]. *There is a voting rule f such that: I) $f(\boldsymbol{L})$ can be computed in polynomial time; II)* **MANIPULATION**$_f$ *is an NP-Complete problem.*

The original proof in Bartholdi et al. [1989] used a variation of Copeland, and similar hardness results hold for common voting rules such as STV [Bartholdi and Orlin, 1991].

4.2.1 FEW CANDIDATES AND COALITIONAL MANIPULATIONS

Note that for a fixed number of candidates m, there is a trivial polynomial-time algorithm for computing a manipulation: simply try all $m!$ possible ballots, which is also a fixed number.

The *coalitional manipulation* problem C-MANIPULATION$_f$ is similar to MANIPULATION$_f$, except it gets as input a set of manipulators I rather than a single manipulator (recall we also discussed coalitional manipulations in Section 3.4). In addition, the input includes the weight of every voter $(w_i)_{i \in I \cup V}$. If f is anonymous, then the coalitional manipulation problem can still be solved with brute force: since there are only $m!$ ways a voter can vote, and we do not care which voter is assigned to which action (only how many), there are "only" $\binom{|I|+m!}{m!}$ different ways that the set of manipulators can vote (i.e., still polynomial in $|I| \leq n$).

The coalitional manipulation problem becomes computationally interesting, so to speak, when we allow voters with non-uniform weights. This breaks anonymity, as it now matters not only *how many* voters cast each ballot, but *which* voters exactly. The number of possible combined ballots that I can submit is $(m!)^{|I|}$ (where $|I|$ may be at the same order as n), so even for a fixed m this is a nontrivial problem that cannot be solved by brute force. Indeed, for many voting rules the coalitional manipulation problem turns out to be NP-hard even for small values of m [Conitzer et al., 2007].

We bring here one such (simple) proof for the Copeland rule as a demonstration. The Copeland winner is the candidate with the highest score $s(c)$, breaking ties lexicographically. Clearly, computing the winner of Copeland is easy—just sum the total weight of voters that support each c vs. any other c'.

Theorem 4.8 **[Conitzer et al., 2007].** *C-MANIPULATION$_{Copeland}$ is an NP-hard problem, even for 4 candidates.*

Proof. We use a reduction from PARTITION, which is known to be NP-hard [Karp, 1972]. In an instance of PARTITION, there is a set of k elements with weights $w_1, \ldots, w_k$, and we should answer whether there is a subset $S \subseteq [k]$ such that $\sum_{i \in S} w_i = \sum_{i \notin S} w_i$. We need to show that given any such instance, we can construct an instance of the MANIPULATION$_{Copeland}$ problem with four candidates and show that a manipulation exists if and only if a valid partition of elements exists.

We denote the candidates by $A = \{a, b, c, p\}$. The set of manipulators I contains one voter with weight w_i for every element of $[k]$. Denote $\overline{w} = \sum_{i \leq k} w_i$. The set V contains three additional pairs of voters, each with weight $\overline{w}$. Their preferences are:

L_1	$p \succ a \succ b \succ c$
L_2	$p \succ b \succ a \succ c$
L_3	$a \succ b \succ c \succ p$
L_4	$b \succ a \succ c \succ p$
L_5	$c \succ p \succ a \succ b$
L_6	$c \succ p \succ b \succ a$

We argue that the set I has a manipulation for p if and only if a valid partition exists. Indeed, note that regardless of how voters in I vote, there is a majority of at least $4\overline{w}$ (out of a total weight of $7\overline{w}$) for the relations: $a \succ c, b \succ c, c \succ p, p \succ a, p \succ b$. In fact, the only relation that the voters in I can affect is between a and b.

Denote the Copland score of a candidate x by $s(x)$. Suppose a valid partition $S, [k] \setminus S$ of weights exists. Then all voters $i \in S$ will rank $a \succ_i b$ and the other voters in $I \setminus S$ will rank $b \succ_i a$. We complete the rest of their votes arbitrarily. Since a and b are now tied, we have $s(a) = s(b) = s(c) = 1$, whereas $s(p) = 2$, and thus p wins.

On the other hand, if a valid partition does not exist, then in any way the voters of I vote, either a beats b or vice versa. Suppose w.l.o.g. that a beats b, then $s(a) = 2 = s(p)$. By lexicographic tie-breaking, p does not win. □

A particular problem of interest that has been open for several years was the complexity of Borda manipulation with $n = 2$ voters (and any number of candidates). This was shown in be an NP-hard problem in Betzler et al. [2011], Davies et al. [2011].

A recent survey of which common voting rules are hard to manipulate appears in Brandt et al. [2016], Section 6.4. We should note that the fact that a problem is NP-hard only means that *some instances* of the problem are hard to solve. It is quite possible that in almost all instances, finding whether a manipulation exists is easy, even in rules like STV. For a detailed discussion of the computational approach to manipulation, see Faliszewski and Procaccia [2010].

4.3 RANDOMIZED VOTING RULES

Suppose we allow our voting rule to flip coins, in other words, return different outcomes with certain probabilities. More formally, an ordinal randomized voting rule is a function that maps any profile $\boldsymbol{L}$ to a lottery (probability distribution) $\sigma \in \Delta(A)$. We denote by $f(\boldsymbol{L}, c) = \sigma(c) \in [0, 1]$ the probability that f returns $c \in A$ on profile $\boldsymbol{L}$. Note that we can also think about f as a lottery over deterministic voting rules.

It is easy to see that we can find randomized rules that violate the Gibbard-Satterthwaite conditions. For example, we can think of a rule that returns any candidate with equal probability, regardless of the profile.

We can also consider somewhat more sophisticated rules, for example, selecting one voter at random, and then use this voter as a dictator. Or, alternatively, select two candidates at random, and then hold a Majority vote among them.

Manipulation of randomized voting rules As with irresolute voting rules (Section 3.5), in order to get a well-defined notion of manipulation, we need to extend preferences over alternatives to *preferences over lotteries*. We do so by assuming cardinal preferences.

Suppose that each voter is endowed with strict cardinal preferences $U_i \in \mathcal{U}_L(A)$ rather than just ordinal ones. Cardinal preferences naturally extend to a complete weak preference order over lotteries by taking the expected utility from each lottery. That is, if some lottery σ selects each candidate $c \in A$ w.p. σ_c, then

$$U_i(\sigma) \triangleq E_{c \sim \sigma}[U_i(c)] = \sum_{c \in A} \sigma_c U_i(c).$$

We will follow this approach in most places in the book where randomization is involved.

We extend the definition of a randomized voting rule above to *cardinal* voting rules, which accept a strict cardinal preference profile $\boldsymbol{U} \in \mathcal{U}_L(A)^n$ as their input. Clearly, any cardinal voting rule induces a unique ordinal voting rule.

Definition 4.9 (Strategyproofness in expectation). A cardinal randomized voting rule f is *strategyproof in expectation* if for any cardinal profile $\boldsymbol{U} \in \mathcal{U}_L(A)^n$, any voter $j \in N$, and any alternative preferences $U'_j \in \mathcal{U}_L(A)$,

$$U_j(f(\boldsymbol{U})) \geq U_j(f(\boldsymbol{U}_{-i}, U'_i)).$$

Definition 4.9 also applies to ordinal voting rules by replacing the input of f with the unique ordinal profile $\boldsymbol{L}$ induced by $\boldsymbol{U}$. A stronger requirement is strategyproofness *ex-post*, which means that for any profile $\boldsymbol{U}$ and any realization (any coin flip), reporting U_i (or L_i in ordinal rules) remains a best reply for every agent i. To demonstrate the difference, consider a vote L'_i that increases the utility of i by 0.6, or decreases the utility by 0.4 with equal probabilities in some profile $\boldsymbol{U}$. Then vote L'_i is a manipulation in expectation, but not ex-post.

Both random dictator and random duple are clearly strategyproof ex-post, and thus also strategyproof in expectation. Unless explicitly stated otherwise, the term *strategyproofness* throughout this book means strategyproofness in expectation.

4.3.1 GIBBARD'S CHARACTERIZATION

It may seem that we might find rules that are strategyproof in expectation, but not ex-post. It turns out, however, that the options we suggested above are roughly the *only* randomized rules that are strategyproof, even in expectation.

Definition 4.10 (Unilateral). A voting rule is *unilateral* if there is some $j \in N$ s.t. for all $U_j, \boldsymbol{U}_{-j}, \boldsymbol{U}'_{-j}$, it holds that $f(\boldsymbol{U}_{-j}, U_j) = f(\boldsymbol{U}'_{-j}, U_j)$.

That is, the outcome is set exclusively based on the input from j.

It is not hard to see that among the unilateral rules, some are strategyproof (e.g., dictatorial rules), while others are not (e.g., selecting j's top choice w.p. $1/3$, and j's last choice w.p. $2/3$). In fact, an ordinal unilateral rule is strategyproof if and only if it is monotone (i.e., selects higher-ranked alternatives with weakly higher probability), see Exercise 4.5(5).

Theorem 4.11 [Gibbard, 1977]. *An ordinal (randomized) voting rule f is strategyproof if and only if it is a lottery over duples and strategyproof unilateral rules.*

In other words, allowing randomization expands the set of strategyproof voting rules, but it is arguable to what extent these rules are "reasonable." One may see Gibbard's theorem as a negative result that extends the G-S theorem to randomized voting rules, or as a positive result specifying many more strategyproof rules. In Section 4.3.3, we consider how much power such rules have in approximating common voting rules. Later in Chapter 5 we will consider randomized strategyproof rules again in the context of mechanism design.

We emphasize that while our definition of strategyproofness assumes cardinal utilities, the voting rules themselves are restricted to (strict) ordinal preferences. Variations of Gibbard's theorem for voting rules whose input is *cardinal preferences* (i.e., with a continuous set of strategies) have also been studied [Barberà et al., 1998, Dutta et al., 2007, Hylland, 1980]. Note that such mechanisms offer more flexibility, for example, the "dictator" may select among several fixed lotteries.

Randomization with dichotomous preferences If we assume that voters have dichotomous weak preferences (see Section 4.1.3), then there is more flexibility in choosing the voting rule. For example, the *random priority* mechanism selects an order over voters uniformly at random and lets each voter eliminate some subset of alternatives. The random priority mechanism is analyzed in Bogomolnaia et al. [2005], which also suggest other mechanisms that are non-truthful but more fair.

4.3.2 STRONGER IMPOSSIBILITY RESULTS

The reader may have noted that the definition of strategyproofness used in Gibbard's theorem is remarkably strong and requires that no voter will have a manipulation in *any* cardinal utility profile. One may adopt a more lax approach similar to the one in the G-S theorem and require only that for every *linear preference profile* $\boldsymbol{L} \in \mathcal{L}^n(A)$; no voter i should be able to manipulate *all* cardinal utility functions U_i that fit L_i [Bogomolnaia and Moulin, 2001, Postlewaite and Schmeidler, 1986]. Equivalently, this means that no voter has an *SD-manipulation*: a vote whose outcome *stochastically dominates* his truthful vote.

Since this alternative definition of strategyproofness (henceforth, *SD-strategyproofness*) forbids a smaller set of manipulations than strategyproofness in expectation,[3] a long-standing open question (made explicit in Aziz et al. [2013]) had been whether it allows for a larger set of SD-strategyproof voting rules.

One variation of this question was settled in a very recent paper by Brandl et al. [2018], who showed the following by making a nontrivial use of computer programs (SMT solvers).

Theorem 4.12 [Brandl et al., 2018]. *For $m \geq 4$ and $n \geq 4$, no anonymous, neutral, and Pareto-efficient (randomized) voting rule is SD-strategyproof.*

Note that the theorem does not strictly generalize Theorem 4.11, and the question is still open if we relax the requirements of anonymity and neutrality.

One should note that many characterization results of randomized strategyproof mechanisms in various domains—some of which appear later in this book—rely in some way on Gibbard's theorem. It is quite possible that some of these results could be strengthened as well by using Theorem 4.12 instead.

[3] One can also compare SD-strategyproofness to TBD-strategyproofness (Definition 3.25). TBD-manipulation allows any cardinal utility profile (and is more permissive than SD-manipulation in that respect), but requires the voter to benefit in any randomized tie-breaking.

4.3.3 OUTPUT APPROXIMATION

Some randomized voting rules have been introduced as a strategyproof randomized versions of common deterministic rules. This includes in particular the *probabilistic Borda* rule, which selects a candidate at random according to its Borda score [Heckelman, 2003].

However, a more structured approach is to use the target rule as an optimization goal, and find a (randomized) strategyproof that best approximates this goal. Consider a rule with a natural notion of candidate score $s(a)$, such as Plurality, Borda, or Maximin. For any alternative $a \in A$ we can compare $s(a)$ to the score of the "true" winner of the voting rule $s(f(\boldsymbol{L}))$.

Definition 4.13 (Output approximation). A (randomized) voting rule f is a γ *output-approximation* of a scoring rule g if for any profile $\boldsymbol{L}$, $E[s_g(f(\boldsymbol{L}))] \geq \gamma \cdot s_g(g(\boldsymbol{L}))$, where s_g is the scoring used by rule g.

Procaccia [2010] suggested to limit ourselves to voting rules that are strategyproof (i.e., fall under Gibbard [1977] characterization) and within this set select a rule that approximates our target rule g as well as possible.

For PSRs, this is done by selecting one voter uniformly at random, and then define a particular lottery over unilateral rules that depends on the scoring function α (e.g., for Plurality, this always selects the most preferred candidate of this voter).

Theorem 4.14 Procaccia **[2010].** *Let g be any PSR, then there is a randomized strategyproof voting rule f such that f is a $\Omega(\frac{1}{\sqrt{m}})$ output-approximation of g. This bound is tight for the Plurality rule.*

That is, for a fixed number of candidates m, we get a fixed approximation ratio regardless of the number of voters.

Proof for Plurality. For $g = f^{PL}$, the randomized voting rule f we will use is the random dictator. Denote by $s_j = s_g(j)$ the number of voters ranking j first, and assume w.l.o.g. that candidate 1 is the winner, which means $s_1 \geq n/m$. We divide into two cases.

Suppose first that $s_1 > \frac{n}{\sqrt{m}}$. Then the expected score of f is

$$E[s_g(f(\boldsymbol{L}))] = \sum_{j \in A} Pr_{f(\boldsymbol{L})}[j] \cdot s_j = \sum_{j \in A} (s_j)^2/n \geq (s_1)^2/n > s_1 \frac{n}{n\sqrt{m}} = \frac{1}{\sqrt{m}} s_g(g(\boldsymbol{L})).$$

If $s_1 \le \frac{n}{\sqrt{m}}$, then let $s = \sum_{j \neq 1} s_j = 1 - s_1$.

$$\begin{aligned}
E[s_g(f(\boldsymbol{L}))] &= \sum_{j \in A} (s_j)^2/n \ge \frac{1}{n} \sum_{j \neq 1} (s_j)^2 = \frac{m-1}{n} \left(\frac{1}{m-1} \sum_{j \neq 1} (s_j)^2 \right) \\
&\ge \frac{m-1}{n} \left(\frac{1}{m-1} \sum_{j \neq 1} s_j \right)^2 = \frac{1}{n(m-1)} s^2 && \text{(Jensen inequality)} \\
&> \frac{1}{n(m-1)} \left(\frac{n}{2} \right)^2 = \frac{n}{4(m-1)} > \frac{n}{4m} && \text{(since } s > 1 - s_1 > \tfrac{n}{2}) \\
&\ge \frac{1}{4\sqrt{m}} \frac{n}{\sqrt{m}} \ge \frac{1}{4\sqrt{m}} s_1 = \frac{1}{4\sqrt{m}} s_g(g(\boldsymbol{L})),
\end{aligned}$$

as required. □

More generally, the randomized rule f used in Procaccia [2010] for a PSR with scoring vector α, is a random unilateral rule, where each voter i is selected with uniform probability, and each alternative $j \in A$ is selected with probability $\frac{1}{\sum_{j'} \alpha_{j'}} \alpha_{L_i^{-1}(j)}$. Note that for Plurality, we get random dictator as a special case since $\alpha_{Plurality} = (1, 0, 0, \ldots, 0)$. Incidentally, by applying this procedure to the Borda rule, we get the probabilistic Borda rule of Heckelman [2003]. The approximation factor we get for Borda is a constant: better than 0.5.

Notably, this approach requires the weakest assumptions on voters' behavior—we only assume they follow their weakly dominant strategy to be truthful. On the other hand, we get an outcome that is only an approximation. Further, this is approximation in expectation, and in a particular realization the outcome may be much worse than g.

4.4 ALMOST-STRATEGYPROOF RULES

4.4.1 APPROXIMATION WITH ALMOST-STRATEGYPROOF RULES

Núnez and Pivato [2016] combined the frequency of manipulation analysis (covered in Section 3.3) with Gibbard's randomized strategyproof rules. Given any deterministic voting rule g (the "target rule"), the randomized voting rule $f_{g,\varepsilon}$ works as follows on any profile $\boldsymbol{L}$:

- with probability $1 - \varepsilon$, return $g(\boldsymbol{L})$;
- otherwise, select a voter $i \in N$ uniformly at random and a pair of candidates $(a, b) \in A$ uniformly at random, and return the one more preferred by i.

Intuitively, $f_{g,\varepsilon}$ is almost-strategyproof because the probability that g can be manipulated in a random profile $\boldsymbol{L}$ (and hence the expected utility gain from manipulation) is low, and in particular lower than the (expected) loss from manipulation when i is selected. On the other hand, $f_{g,\varepsilon}$ and g are close in a similar sense to Definition 3.15, especially when n is large.

Núnez and Pivato define f and g to be *asymptotically equivalent* if $C(f, g) \overset{n\to\infty}{\to} 0$. Their definitions and results take into account various distributions of profiles, and not necessarily the impartial culture. Then they show that for every deterministic voting rule g and $n \in \mathbb{N}$ there is some sequence of constants $\varepsilon(n)$ such that g and $f_{g,\varepsilon(n)}$ are asymptotically equivalent and $f_{g,\varepsilon(n)}$ is asymptotically truthful.

4.4.2 DIFFERENTIAL PRIVACY

Consider a randomized algorithm f that operates on some data $\boldsymbol{D}$ and returns some outcome $a \in A$. We can think of $\boldsymbol{D}$ as a table with one row per person, and where columns specify various information about this person.

An algorithm f is *differentially private* if for all $\boldsymbol{D}, \boldsymbol{D}'$ that differ by one person, and all $S \subseteq A$ we have $Pr[f(\boldsymbol{D}) \in S] \cong Pr[f(\boldsymbol{D}') \in S]$. That is, we cannot tell by the outcome (almost) anything about a particular person in the data.

There is abundant literature on constructing differentially private algorithms, or modifying existing algorithms so as to make them differentially private. This is typically by adding noise to the dataset in a particular way, see for example Dwork and Lei [2009].

Differentially private voting rules Intuitively, a (randomized) voting rule f is differentially private if for all profiles $\boldsymbol{L}, \boldsymbol{L}'$ that differ by a single voter, and for any subset of alternatives $S \subseteq A$, $Pr[f(\boldsymbol{L}) \in S] \cong Pr[f(\boldsymbol{L}') \in S]$. That is, a single voter has very small influence on the outcome. While it seems that this is the case in most voting rules, note that in standard deterministic rules (like Plurality), there are very few profiles in which a voter is pivotal, but in those profiles the voter has large influence (changes the outcome with certainty). In contrast, in differentially private rules, in *every* profile a single voter has only negligible influence.

How does this relate to strategyproofness? Since the influence of every voter is small, the (expected) gain from manipulation is also small. Thus the main idea is to take some deterministic voting rule f and modify it by adding some noise to the votes before computing the outcome. If done properly, this will have very little effect on the outcome in most profiles while eliminating the situations where a voter has a strong incentive to manipulate. We slightly modify Definition 4.9:

Definition 4.15 (ε-strategyproofness). A cardinal (randomized) voting rule f is ε-strategyproof if for any cardinal profile $\boldsymbol{U} \in \mathcal{U}_L(A)^n$, any voter $j \in N$, and any alternative utility scale $U'_j \in \mathcal{U}_L(A)$,

$$U_j(f(\boldsymbol{U})) + \varepsilon > U_j(f(\boldsymbol{U}_{-j}, U'_j))$$

Thus, strategyproof rules are 0-strategyproof. As in Section 4.3, we focus on ordinal voting rules that ignore the cardinal information in the reported preferences. Note that ε-strategyproofness differs from the notion of manipulation power we saw in Section 3.3. There, we looked at the expected gain (the probability of manipulation) over all profiles, whereas here even a single profile can be a witness that a voting rule is far from strategyproofness.

Let $d(\boldsymbol{L}, \boldsymbol{L}') = \sum_{i \in N} [\![L_i \neq L'_i]\!]$ count the number of voters that vote differently in $\boldsymbol{L}, \boldsymbol{L}'$. Note that $d : \mathcal{L}(A) \times \mathcal{L}(A) \to \mathbb{N}$ is a pseudometric.

Definition 4.16 (Input approximation). A randomized voting rule f is a δ input-approximation of another voting rule g, if for any profile $\boldsymbol{L}_f$ and every coin toss, there is another profile $\boldsymbol{L}_g$ such that:

- $d(\boldsymbol{L}_f, \boldsymbol{L}_g) \leq \delta \cdot n$ and
- $f(\boldsymbol{L}_f) = g(\boldsymbol{L}_g)$.

That is, for every $\boldsymbol{L}$ and every realized outcome of f it is possible that only δ votes would change so that the target rule g would result in the same outcome. Note that this is a very different notion of approximation than the one used in Section 4.3.3. As a more concrete example, consider the Plurality rule $g = f^{PL}$, and define f by first sampling 80 random voters and replace their vote with an arbitrary candidate, and then compute the Plurality winner. Then f is a $\frac{80}{n}$ input-approximation of g. On the other hand, a rule f' that with probability 0.1 selects a dictator (and otherwise uses g) is not a 0.1 input-approximation of g. This is because in the case a dictator is used and votes for a, and all other voters vote for b (in $\boldsymbol{L}_f$), we have to change the vote of almost $n/2$ voters in $\boldsymbol{L}_g$ to get $g(\boldsymbol{L}_g) = a$.

Theorem 4.17 [Birrell and Pass, 2011]. *For any deterministic voting rule g over m candidates, and any $\varepsilon > 0$, there is a randomized voting rule $f_{g,\varepsilon}$ such that:*

- *$f_{g,\varepsilon}$ is ε-strategyproof and*
- *$f_{g,\varepsilon}$ is a δ input-approximation of g, for $\delta = O(\frac{m^2}{\varepsilon})$.*

In other words, we can construct a voting rule f where the incentive to manipulate is as low as we want. This voting rule will coincide with g except in "borderline" profiles where a small number of voters could change the outcome. The size of this margin is inversely proportional to the incentive to manipulate. Birrell and Pass also show a lower bound of $\Omega(\frac{1}{nm^3})$ on the ε-strategyproofness of a nontrivial rule (i.e., not a mixture of duples and dictatorial rules), thereby provide yet another quantified relaxation of Gibbard [1977] theorem.

The proof of Theorem 4.17 (and other, more nuanced results in the paper) is mathematically sophisticated and quite long, and we do not bring it here. Instead, we can discuss the underlying assumptions. In reality, voters have some uncertainty over the outcome even without adding noise, and thus they are already "differentially private" to some extent from the perspective of the voter. Yet, voters do manipulate even though their expected gain is negligible, which raises the question if ε-strategyproofness is a good enough guarantee in practice.

4.5 EXERCISES

1. Consider the following profile. Is there an order $\mathcal{O}$ such that all six voters are single-peaked w.r.t. $\mathcal{O}$? explain. Remove one voter and find an order such that the remaining profile is single-peaked. Who is the median voter? What is the median candidate?

L_1	$A \succ E \succ C \succ B \succ D$
L_2	$A \succ E \succ C \succ B \succ D$
L_3	$E \succ A \succ D \succ C \succ B$
L_4	$E \succ C \succ B \succ D \succ A$
L_5	$C \succ B \succ E \succ D \succ A$
L_6	$D \succ B \succ C \succ E \succ A$

2. Given a tree $\mathcal{T}$ over nodes A (undirected, unweighted), we say that R_i is single-peaked w.r.t. $\mathcal{T}$ if there is some "peak candidate" $a_i^* \in A$ s.t. candidates that are closer to a_i^* along a branch are more preferred. Formally, for all $x, y \in A$, if x is on the unique path between a_i^* (included) and y (excluded), then $a_i^* \succsim_i x \succ_i y$. The median of a subset of nodes $B \subseteq A$ over a tree $\mathcal{T}$ is a node $a^* \in A$ such that each subtree of the tree rooted by a^* contains strictly fewer than $|B|/2$ points from B. Generalize the Median mechanism to preference profiles that are single-peaked w.r.t. a tree and prove that it is strategyproof.

3. Consider $A = \{0, 1\}^2$. Find a natural set of properties $\boldsymbol{P}$ for A (hint: $|\boldsymbol{P}| = 4$). When is a candidate x between candidates y and z? Give examples to preferences that are single-peaked and non-single-peaked w.r.t. the properties you defined. For an odd n and any property P, let $\mathcal{W}_P$ be all coalitions of size $> n/2$. What is the outcome of the Voting-by-Issues mechanism?

4. The voting rule f first checks if there is a Condorcet winner, and otherwise selects the Veto winner (break ties lexicographically). Show that it is easy to decide if a given coalition $I \subseteq N$ has a group manipulation in favor of candidate a.

5. Any ordinal unilateral randomized rule f corresponds to a probability distribution p_f over $[m]$, where $p_f(k)$ is the probability that $L_i(k)$ is selected. Prove that f is strategyproof if and only if p_f is non-increasing (i.e., candidates ranked higher are selected with weakly larger probability).

6. Voting rule f for 3 candidates replaces 1,000 of the votes with 1,000 random votes, and then runs the Plurality rule. Show that f is 0.04-strategyproof.

7. Suppose that $n = 9$, and consider the median-of-medians rule which first computes the median of each group $i \in \{1, 2, 3\}, j \in \{4, 5, 6\}, k \in \{7, 8, 9\}$, and then computes the median of $\{i, j, k\}$. Consider the properties P_{lx}, P_{rx} defined in Section 4.1.2. What are the winning coalitions W_P?

Show that $f(\boldsymbol{L})$ is always in $F_P(\boldsymbol{L})$. Conclude that both voting rules above are strategyproof.

Consider the same question with $n = 8$ voters, where instead of l_9 we use the fixed location x^*.

CHAPTER 5

Voting and Mechanism Design

Throughout this book, the incentives of the voters are pretty straightforward, as each voter is interested in the election of the most preferred candidate/alternative. In contrast, there is no agreed notion in the social choice literature for what is a socially good outcome. To quote Downs [1957, p. 136]:

> ...it is not clear what is meant by "social welfare," nor is there any agreement about how to "maximize" it.

This is partly due to the fact that voters are often assumed to have ordinal preferences, and any voting rule can be considered as a different definition of optimal aggregation. There are many axiomatic, statistical, topological, and other arguments in favor or against each rule, even when assuming all voters are truthful (see Brandt et al. [2016] for further discussion).

In contrast to the common abstract model of voting, in many specific situations there are both well-defined cardinal utilities for each alternative and a clear social goal: for example, to maximize the sum of utilities. Without further constraints, one could simply define a voting rule that asks each agent for her type and select the alternative maximizing the social goal. However, such a voting rule would clearly not be truthful.

The general definition of a mechanism we will use is a (possibly randomized) function $g : \mathcal{U}(A)^n \to A$ from cardinal preference profiles to an outcome, possibly with constraints on the input or output, and/or other modifications.

We say that mechanism g is *ordinal* if it only uses the ordinal information in $\boldsymbol{U}$. That is, if for every $\boldsymbol{U}, \boldsymbol{U}'$ that fit the same ordinal preferences $\boldsymbol{L}$, $g(\boldsymbol{U}) = g(\boldsymbol{U}')$. We assume in this chapter that a voter may be indifferent between alternatives, except in Section 5.2 where we restrict preferences to $\mathcal{U}_L(A)$ (as in Section 4.3).

Mechanism design aims at designing mechanisms that are strategyproof and yet do well in achieving some optimization goal [Nisan, 2007]. In this chapter we focus on utilitarian social welfare (sum of utilities), but other common goals are egalitarian welfare (lowest utility) or the designer's revenue.

We can also think of course on mechanisms that are not strategyproof but guarantee good outcomes *in equilibrium*. We discuss such results in Part II of the book, and in particular in the context of implementation theory (Section 6.2).

Welfare and approximation Assuming agent i's utility for each outcome $c \in A$ is given by $U_i(c)$, the *social welfare* of outcome c in profile $\boldsymbol{U}$ is $SW(c, \boldsymbol{U}) = \sum_{i \in N} U_i(c)$. We omit the parameter $\boldsymbol{U}$ when clear from the context. We denote by $c^*(\boldsymbol{U}) \in \text{argmax}_{c\in A} \sum_{i\in N} U_i(c)$ (or sometimes just c^*) an arbitrary candidate that maximizes the social welfare, and by $SW^* = SW(c^*(\boldsymbol{U}), \boldsymbol{U}) = \max_{c\in A} \sum_{i\in N} U_i(c)$ the maximal social welfare in profile $\boldsymbol{U}$.

Definition 5.1 (Welfare approximation). A (possibly randomized) mechanism g is γ *welfare-approximation* if for any profile $\boldsymbol{U}$, $E[SW(g(\boldsymbol{U}))] \geq \gamma \cdot SW^*(\boldsymbol{U})$, where expectation is taken over randomization of the mechanism, if relevant.

In contrast to output approximation and input approximation of a specific "target voting rule" f (in Sections 4.3 and 4.4, respectively), the optimal welfare is inferred directly from the profile $\boldsymbol{U}$.

Often when we write "mechanism" we mean a family of mechanisms, one for every possible input size (n and m). Thus the approximation ratio of (a family of) mechanisms may be a function of n and m.

Outline The literature on mechanism design is immense, and a full coverage is completely outside the scope of this book. This chapter overviews several models and results in mechanism design that are strongly related to social choice and voting.

There are several textbooks on mechanism design *with money*. One extensive text that focuses on approximation is based on Jason Hartline's lecture notes [Hartline, 2013]. In Section 5.1 we show one fundamental result from this literature: if monetary transfers are allowed, then there is a truthful mechanism (namely, the VCG mechanism) that maximizes the social welfare.

However, in many domains the use of monetary payments is forbidden, impractical, unethical or otherwise undesirable [Rothkopf, 2007], in which case a mechanism is simply a function (game form) that accepts inputs from agents, and outputs an outcome from a pre-defined set. The literature on mechanism design *without money* is more sparse, and papers are typically dedicated to a specific domain (matching, facility location, fair allocation, etc.).

We show the application of the mechanism design approach to general social choice problems (Section 5.2), to facility location problems, where voters prefer to be as close as possible to the facility (Section 5.3), and to judgment aggregation problems, where there are logical constraints on the possible outcomes (Section 5.4). In each section we show a full or partial characterization of strategyproof mechanisms, and summarize known results regarding upper and lower bounds on the best possible approximation ratio of such mechanisms.

5.1 PAYMENTS

Adding payments allows us to transfer utility between agents with much flexibility, thereby aligning their incentives. The main tool we apply is a powerful mechanism with many applications in mechanism design, called the Vickrey–Clarke–Groves mechanism (VCG). We should also mention *quadratic voting* [Lalley and Weyl, 2014], which is a more recent tool for aligning the preferences of the society over $m = 2$ alternatives, and that will not be described here.

Voting with money To be able to compare voting outcomes and money, we assume voters have cardinal utilities over candidates as in Section 4.3. That is, $U_j(c) \in \mathbb{R}$ is the utility that voter j assigns to candidate c. We now reformulate the voting process as an auction mechanism, where each voter announces her value for each candidate. We can think of these values as "bids," and a voter may bid higher or lower than her true value.

We use the following formal definition of a mechanism with payments (in this section simply referred to as a mechanism).

Definition 5.2 (Mechanism with payments [Nisan, 2007]). A (direct revelation) *mechanism with payments* is a tuple $\langle g, \boldsymbol{p} \rangle$, where $g : \mathcal{U}(A)^n \to A$ is an SCF; and $\boldsymbol{p} = (p_1, \ldots, p_n)$ is a vector of payment functions of the form $p_i : \mathcal{U}(A)^n \to \mathbb{R}$.

That is, agents report their (cardinal) preferences $U_1, \ldots, U_n$; the outcome $c = g(\boldsymbol{U})$ is selected; and each agent i pays $p_i = p_i(\boldsymbol{U})$.

The utility of agent i is then $U_i(c) - p_i$, that is, linear in the paid amount, and we assume that each agent tries to maximize $U_i(c) - p_i$. This form of utility is known as *quasi-linear utility*.

Manipulations Note that as with any other game form, any mechanism $\langle g, \boldsymbol{p} \rangle$ together with (cardinal) utility functions $\boldsymbol{U}$ induce a game. The mechanism $\langle g, \boldsymbol{p} \rangle$ is *strategyproof* if in all such games, reporting one's true type U_i is a dominant strategy for every voter.

The logical thing to do given the reported values, is to select a candidate that maximizes the social welfare, in other words, set $g(\boldsymbol{U}) = c^*(\boldsymbol{U})$. We then need to decide the amount we charge each agent.

As a first attempt, we can charge each voter j the amount of her bid, that is, $p_j = U_j(c^*)$. However, it is not hard to see that this is not strategyproof. Indeed, suppose all voters assign value 1 to c^* and 0 to all other candidates. Then j can manipulate by reporting $U'_j(c^*) = 0$: c^* will still be selected, the payment of j will drop, and her utility will strictly increase.

5.1.1 THE VCG MECHANISM

The Vickrey–Clarke–Groves (VCG) mechanism [Clarke, 1971, Groves, 1973, Vickrey, 1961] collects all information from the agents. Then it computes the optimal outcome and charges each agent i the "damage" that this agent inflicts upon the other agents, that is, the difference between the maximal social welfare in a world where i does not exist, and the maximal social welfare (of all except i) in the current world. For a thorough exposition of VCG and a range of applications, see Nisan [2007]. Here we focus on the connections to social choice.

While the VCG mechanism guarantees truthfulness in a wide variety of applications, it turns out that as a voting rule it has a particularly attractive form (see Algorithm 5.1): agent j only pays *if she is pivotal*, and in that case only pays the minimal bid that would still make c^* win.

By definition, VCG always selects the candidate maximizing the social welfare. See Figure 5.1 (left) for a graphical demonstration.

Algorithm 5.1 The VCG mechanism

Input: reported utilities $\hat{U}_i(c)$ for all $i \in N, c \in A$
Let $c^* = \text{argmax}_{c \in A} \sum_{i \in N} \hat{U}_i(c)$
for each $j \in N$ {
  Compute $\widehat{SW}_{N \setminus \{j\}} = \max_{c \in A} \sum_{i \in N \setminus \{j\}} \hat{U}_i(c)$ //the social welfare when j does not exist
  Compute $\widehat{SW}_N^{-j} = \sum_{i \in N \setminus \{j\}} \hat{U}_i(c^*)$ //the current social welfare without j
  Set $p_j = \widehat{SW}_{N \setminus \{j\}} - \widehat{SW}_N^{-j}$
}

Theorem 5.3 **[Clarke, 1971].** *The VCG voting rule is strategyproof. That is, for any* $\mathbf{U} \in \mathcal{U}(A)^n$*, it is a dominant strategy for each agent i to report her true type* $\hat{U}_i = U_i$ *under the VCG mechanism.*

Proof. The crucial observation is that $\widehat{SW}_{N \setminus \{i\}} = SW_{N \setminus \{i\}}$ regardless of how agent i reports, since it is computed only from the reports of the other agents. If agent i reports truthfully $\hat{U}_i = U_i$, then c^* is selected and the social welfare is

$$\begin{aligned} SW_N &= SW_N^{-i} + U_i(c^*) = \widehat{SW}_N^{-i} + U_i(c^*) \\ &= (\widehat{SW}_{N \setminus \{i\}} - p_i) + U_i(c^*) = SW_{N \setminus \{i\}} + (U_i(c^*) - p_i). \end{aligned}$$

Suppose agent i reports $\hat{U}_i = U'_i$, then some (possibly) other candidate c' is selected, and the social welfare is

$$\begin{aligned} SW'_N &= SW'^{-i}_N + U_i(c') = \widehat{SW}'^{-i}_N + U_i(c') \\ &= (\widehat{SW}'_{N \setminus \{i\}} - p'_i) + U_i(c') = SW_{N \setminus \{i\}} + (U_i(c') - p'_i). \end{aligned}$$

Since by definition, $SW_N \geq SW'_N$, we have that the utility (including payment) of i for truthful reporting, $U_i(c^*) - p_i$ is weakly higher than the utility when reporting U'_i, which is $U_i(c') - p'_i$. □

A less formal intuition is as follows: a voter i can try to manipulate in one of two ways. She can try to reduce her bid on c^*. Note that p_i (and thus the utility of i) will not change unless the winner changes. But if the winner changes to c'', then $U_i(c^*) - p_i \geq U_i(c'') - p'_i$ (see Figure 5.1), which means i does not gain. Likewise, in order to change the winner by increasing the bid on c'', i must increase her bid enough so that the new payment p'_i is greater that all of $U_i(c'')$.

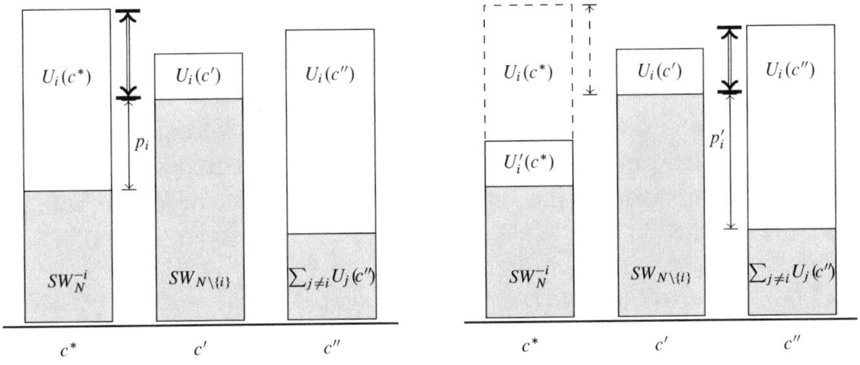

Figure 5.1: Left: A graphical demonstration of the VCG mechanism. The gray bar indicates the total utility voters other than i attribute to each candidate. We can see that the payment p_i is independent of U_i. The double arrow indicates the utility of voter i. Right: An attempted manipulation of voter i, by reporting a lower utility $U_i'(c^*)$ on candidate c^*. The new winner c'' has higher utility but also higher payment p_i', so that i does not gain.

Extensions Once payments are allowed, VCG is not the only strategyproof mechanism that exists. After requiring some additional natural properties, Roberts [1979] characterizes all mechanisms that are strategyproof for voters with arbitrary cardinal utilities. These mechanisms (called *affine maximizers*) all look like weighted variations of VCG and maximize weighted versions of the social welfare.

Both VCG and affine maximizers rely heavily on the assumption that voters' utilities are quasi-linear. However, in many real situations utilities may not be quasi-linear due to wealth effects, liquidity, and other issues [Cramton, 1997, Pratt, 1964]. Recently, Ma et al. [2016, 2018] showed that relaxing the assumption of quasi-linear utilities even slightly leads to further impossibility results in the spirit of the G-S theorem.

In principle, after the voting is complete we could somehow give the money back to the voters. However, doing so in an arbitrary way (e.g., return $\frac{1}{n}$ of the collected amount to each voter) might make the entire mechanism not strategyproof. The literature on *redistribution* studies how at least some of the money can be returned to the voters while maintaining strategyproofness and other properties [Cavallo, 2006, Guo and Conitzer, 2009].

5.2 RANGE VOTING

The first domain on which we will demonstrate the concepts of truthful approximation mechanisms is (admit you did not expect it)—voting!

Range Voting[1] allows voters to express their strict cardinal preferences over candidates ($U_i \in \mathcal{U}_L(A)$, normalized such that $\min_{a \in A} U_i(a) = 0$ and $\max_{a \in A} U_i(a) = 1$), and selects the one maximizing the sum of utilities. In other words, it always returns $c^*(\boldsymbol{U})$.

Even without the G-S theorem, it is obvious that Range Voting is not truthful, as voters in a minority position can always gain by reporting more extreme preferences. Trying to approximate the optimal outcome (i.e., approximate the outcome of Range Voting) with any deterministic rule is futile:

Proposition 5.4 *Any deterministic strategyproof voting rule f has a welfare-approximation ratio of $\gamma \leq \frac{1}{n}$, for any $n \geq 2$ and $m \geq 3$.*

Proof. Let $\varepsilon > 0$ be an arbitrarily small constant. We prove that no f has an approximation ratio better than $\frac{1}{n} + \varepsilon$.

By the G-S theorem, the only strategyproof voting rules are duples (rules with range of size 2) and dictatorial rules. If f is a duple, then its range does not contain some $c \in A$. Thus for the profile where all voters assign $U_i(c) = 1$ and $U_i(c') < \varepsilon$ for all $c' \neq c$, we have $SW(f(\boldsymbol{U})) < \varepsilon n$ and $SW^* = SW(c) = n$ (thus, $\gamma < \varepsilon$).

If f is dictatorial (w.l.o.g. always selects voter 1 as a dictator), then consider a profile where $U_1(a_1) = 1, U_1(a_2) = 1 - \frac{\varepsilon}{n}$ and $U_i(a_1) = 0, U_i(a_2) = 1$ for all $i > 1$. That is, voter 1 is almost indifferent but very weakly prefers candidate a_1. We then have $SW(f(\boldsymbol{U})) = SW(a_1) = 1$, whereas $SW^* = SW(a_2) > n - \frac{\varepsilon}{n}$. Thus $\gamma \leq \frac{SW(a_1)}{SW^*} < \frac{1}{n - \varepsilon/n} < \frac{1}{n} + \varepsilon$. □

On the other hand, it is easy to see that a dictatorial rule has an approximation ratio of at least $\frac{1}{n}$, since for the dictator i, we have $SW(f(\boldsymbol{U})) = SW(\text{top}(U_i)) \geq U_i(\text{top}(U_i)) = 1$, whereas $SW^* \leq n$.

5.2.1 APPROXIMATION BY RANDOMIZED VOTING RULES

An approximation factor that shrinks linearly with the number of voters is not very impressive, and thus deterministic rules are not helpful.

In a fairly recent work, Filos-Ratsikas and Miltersen [2014] suggested to find among the class of randomized strategyproof rules (characterized in Theorem 4.11), the ones that give the best possible approximation for the social welfare. Recall that we already saw in Section 4.3 a similar use of randomized voting rules for approximation. The difference is the goal we are trying to approximate. Output approximation is suitable for voters with ordinal preferences, where the randomized voting rule g was supposed to approximate the outcome of some deterministic target rule f. In contrast, in this section we follow a more standard mechanism design approach by trying to approximate the optimal (cardinal) social welfare. We do note, however, that both approaches rely on the characterization of Gibbard [1977].

[1]See, for example, `http://rangevoting.org/`

The main result of Filos-Ratsikas and Miltersen is a tight-to-a-constant approximation bound that does not depend on n at all and decreases sub-linearly with m. They also prove a tight bound of $\gamma = \frac{2}{3}$ for the case of $m = 3$. While in the paper there is only a mechanism with $\gamma = \Omega(m^{-\frac{3}{4}})$, Filos-Ratsikas closes this gap in his thesis [Filos-Ratsikas, 2015].

We omit the proof of the lower bound, but describe the mechanism that obtains it: the mechanism g^{XL} first selects a voter i uniformly at random. Then, w.p. 0.5 the mechanism returns $\text{top}(L_i)$, and w.p. 0.5 the mechanism returns a candidate selected uniformly at random from the top $\lfloor m^{\frac{1}{3}} \rfloor$ candidates of L_i.

Theorem 5.5 **[Filos-Ratsikas, 2015].** *Mechanism g^{XL} has a welfare approximation ratio of $\gamma = \Omega(m^{-\frac{2}{3}})$ for any n.*

Theorem 5.6 **[Filos-Ratsikas and Miltersen, 2014].** *Any ordinal (randomized) strategyproof voting rule g has an approximation ratio of $\gamma < 5m^{-\frac{2}{3}}$, for any $m \geq 10$ (and sufficiently large n).*

Proof. Let $k = \lfloor m^{1/3} \rfloor$, $t = \lfloor m^{2/3} \rfloor$, and $n = m - 1 + t$. W.l.o.g., the mechanism g is neutral and anonymous (easy exercise). We thus write g as a combination of duples $f_{(a,b)}$ and strategyproof unilateral rules f_i, where any $f_{(a,b)}$ has the same probability, and any f_i has the same probability.

We now describe a single profile for which any such mechanism performs badly. Let $M_1, \ldots, M_t$ be a partition of $\{1, \ldots, kt\}$ with k candidates in each set M_j (note that $kt \leq m$). The bad profile has the following voters.

- For each $i \in \{1, \ldots, m-1\}$, a voter i with utilities $U_i(i) = 1, U_i(m) = 0$, and $U_i(c) < \frac{1}{2m^2}$ for all $c \in A \setminus \{i, m\}$.
- For each $j \in \{1, \ldots, t\}$ a voter $j' = m - 1 + j$ with utilities $U_{j'}(c) > 1 - \frac{1}{m^2}$ for all $c \in M_j$, $U_{j'}(m) = 1 - \frac{1}{m^2}$, and $U_{j'}(c) < \frac{1}{m^2}$ for all other candidates.

Note that $SW^* \geq SW(m) = (1 - \frac{1}{m^2})t$ whereas $SW(c) < 1 + 2m\frac{1}{2m^2} < 1 + \frac{1}{m}$ for any $c \neq m$. Thus, the conditional expected approximation ratio given that the mechanism does not elect m is at most $\frac{1+\frac{1}{m}}{(1-\frac{1}{m^2})t} \leq 2m^{-\frac{2}{3}}$. We therefore only need to bound the probability that candidate m is elected.

The chance that m will be in the range of a random duple is at most $\frac{2}{m}$, as they are selected uniformly at random due to neutrality.

Since a strategyproof ordinal unilateral rule f_i must be monotone (see Exercise 4.5(5)), the probability that f_i assigns to m is at most $\frac{1}{L_i^{-1}(m)}$.

Recall that $m - 1$ voters rank m last (for which $f_i(m) \leq 1/m$), and the other t voters rank m at place $k + 1$ (for which $f_i(m) < \frac{1}{k}$). This means that candidate m is chosen with probability at most

$$\frac{2}{m} + \frac{t}{n}\frac{1}{k} + \frac{m-1}{n}\frac{1}{m} < \frac{3}{m} + \frac{\lfloor m^{2/3} \rfloor}{m \lfloor m^{1/3} \rfloor} < 3m^{-\frac{2}{3}},$$

since $m \geq 10$. We conclude that on the bad profile, the expected approximation ratio of any mechanism in the decomposition is

$$Pr(m)1 + \sum_{c \neq m} Pr(c)\frac{SW(c)}{SW^*} \leq 3m^{-\frac{2}{3}} \cdot 1 + 2m^{-\frac{2}{3}} = 5m^{-\frac{2}{3}}.$$

Therefore, the expected approximation ratio of g on the bad profile is also at most $5m^{-\frac{2}{3}}$. □

Note that the theorem restricts attention to *ordinal* mechanisms, which cannot be optimal even without the strategyproofness requirement [Boutilier et al., 2015]. It is an open question whether a better mechanism that uses cardinal information exists.

5.3 FACILITY LOCATION

Facility location can be thought of as a special case of voting, where the alternatives A are possible locations for a facility. More formally, $\langle \mathcal{X}, d \rangle$ is a metric space where d is a metric over the set $\mathcal{X}$. For example, $\mathcal{X}$ may be an Euclidean space or the vertices of a graph. The set of allowed locations for a facility is $A \subseteq \mathcal{X}$. Each agent is assumed to prefer the facility to be as close as possible to her location, thus instead of reporting her entire utility function U_i, she only needs to report her location $x_i \in \mathcal{X}$. The cost (negative utility) of every alternative $a \in A \subseteq \mathcal{X}$ is exactly the distance $d(x_i, a)$.

While there are many papers on strategic facility location, here we focus on results that are most related to strategic voting either conceptually or technically, many of which are rather recent. We follow the formal framework of Procaccia and Tennenholtz [2009, 2013].[2]

It will be convenient to measure the *social cost* rather than the social welfare. Formally, the (utilitarian) social cost of a facility placed at $a \in A$ in profile $\boldsymbol{x} = (x_1, \ldots, x_n)$ is $SC(a, \boldsymbol{x}) = \sum_{i \in N} d(x_i, a)$.

A facility location mechanism is a function $g : \mathcal{X}^n \to A$, mapping the positions of all n agents to a single winning position. The special case where $A = \mathcal{X}$ is called the *unconstrained case*, as the facility can be placed anywhere, and in particular wherever an agent can be placed. The *cost approximation ratio* of g is the smallest γ s.t. for any input $\boldsymbol{x}$, $E[SC(g(\boldsymbol{x}), \boldsymbol{x})] \leq \gamma \cdot SC^*(\boldsymbol{x})$. Results on the cost approximation ratio of facility location mechanisms are summarized in Table 5.1

Note that many of the properties we defined for voting rules (onto, dictatorial, unilateral, strategyproof, group-strategyproof, etc.) naturally extend to facility location mechanisms.

[2]Facility location games are also closely related to *spatial voting* models, such as the one suggested by Downs where voters elect candidates that are closest to them in some ideological space [Downs, 1957]. However, as these models typically focus on the strategic behavior of *candidates*, they remain outside the scope of this book. See, for example, Hinich [1977] to learn more on the connection between these models.

5.3.1 LOCATION IN A GENERAL METRIC SPACE

While there are many possible aggregation mechanisms, the most natural one g^{opt} would simply take all of the locations $x_1, \dots, x_n$ and return the point $a \in A$ that minimizes the average distance to all input points (in other words, the mechanism always outputs the optimal outcome). Suppose $A = \mathcal{X} = \mathbb{R}^k$ for some dimension k. Clearly, if $g^{opt}(\boldsymbol{x}) \neq x_i$ then agent i can manipulate by reporting a different location x_i' (typically more extreme), thereby moving the facility closer to her real location x_i. This means that the "optimal" mechanism is not strategyproof, at least when $A = \mathcal{X} = \mathbb{R}^k$.

In the constrained version of the problem, it is not hard to create spaces where any preference order L_i over the set A can be attained by appropriately selecting the position $x_i \in \mathcal{X}$. Thus the facility location becomes equivalent to standard voting, as every facility location mechanism g is just a voting rule f_g, where $g(x_1 \dots, x_n) = f_g(L_1, \dots, L_n)$.

We will see below how it follows from the G-S theorem that the only deterministic strategyproof facility location mechanisms for arbitrary spaces are duples or dictatorial. Similarly to Section 5.2, we may ask how good or how bad such mechanisms are, with respect to the social welfare.

Theorem 5.7 **[Meir et al., 2012].** *The mechanism g^1 which selects agent 1 as a dictator has a cost approximation ratio of $2n-1$, and $n-1$ on unconstrained instances.*

Proof. Denote by $a_1^* = g^1(\boldsymbol{x})$ the candidate closest to agent 1 (which is returned by the mechanism), and $d^* = d(x_1, c^*)$. Note that $d(x_1, a_1^*) \leq d^*$. Then

$$
\begin{aligned}
SC(g^1(\boldsymbol{x})) = SC(a_1^*) &= \sum_{i \in N} d(a_1^*, x_i) = d(x_1, a_1^*) + \sum_{i>1} d(a_1^*, x_i) \\
&\leq d^* + \sum_{i>1} (d(a_1^*, x_1) + d(x_1, c^*) + d(c^*, x_i)) \qquad \text{(triangle inequality)} \\
&\leq d^* + \sum_{i>1} (2d^* + d(c^*, x_i)) = (2n-1)d^* + \sum_{i>1} d(c^*, x_i).
\end{aligned}
$$

Denote $D = \sum_{i>1} d(c^*, x_i)$, then $SC(a_1^*) = (2n-1)d^* + D$. For the optimal location c^*, we have $SC(c^*) = \sum_{i \in N} d(c^*, x_i) = D + d(c^*, x_1) = D + d^*$. Thus,

$$
\frac{SC(a_1^*)}{SC(c^*)} \leq \frac{(2n-1)d^* + D}{d^* + D} \leq \frac{(2n-1)d^*}{d^*} = 2n-1,
$$

as required. In the unconstrained case, $d(a_1^*, x_1) = 0$, and thus

$$
SC(a_1^*) \leq \sum_{i>1} (d(x_1, c^*) + d(c^*, x_i)) = (n-1)d^* + D,
$$

and we get the desired bound. □

Theorem 5.8 *Any deterministic strategyproof facility location mechanism*[3] *g has a cost approximation ratio of at least $2n-1$.*

We prove the theorem for $\mathcal{X} = \mathbb{R}^2$. In Exercise 5.5(5) the reader will need to write a similar proof when $\mathcal{X}$ is a circle.

Proof. Let $\varepsilon' > 0$ and $\varepsilon = \varepsilon'/2n$. We define $\mathcal{X} \triangleq \mathbb{R}^2$ and $A \triangleq \{a, b, c\}$ where $a = (0,0), b = (1,0)$ and $c = (0,1)$. We also define $Z \subseteq \mathcal{X}$ as a set of 6 points as follows: $z_{ab} = (\varepsilon, 0), z_{ba} = (0.5+\varepsilon, 0), z_{ac} = (0, \varepsilon), z_{ca} = (0, 0.5+\varepsilon), z_{bc} = (1-\varepsilon, 1), z_{cb} = (1, 1-\varepsilon)$. Note that each such point induces one of the possible 6 preference orders on A. For example, z_{ab} induces the order $a \succ b \succ c$ since $d(z_{ab}, a) = \varepsilon < 0.5 + \varepsilon = d(z_{ab}, b) < 1 < d(z_{ab}, c)$. For every order $L \in \mathcal{L}(A)$ denote by $z_L \in Z$ the point corresponding to this order.

Thus, any facility location mechanism $g : \mathcal{X} \to A$ induces voting rule $f_g : \mathcal{L}(A) \to A$, where $f_g(\boldsymbol{L}) = g(z_{L_1}, \dots, z_{L_n})$. We argue that if g is strategyproof and onto, it must be dictatorial on Z, meaning that there is an agent i such that for all profiles $\boldsymbol{x} \in Z^n$, $g(\boldsymbol{x}) = a_1^*$. Indeed, f_g is onto since g is, and f_g is strategyproof since g is: if there is a manipulation $\boldsymbol{L}, L_i'$ in f_g, then agent i also has a manipulation in profile $z_{L_1}, \dots, z_{L_n}$ in g by reporting $z_{L_i'}$. By the G-S theorem, f_g must be dictatorial. This in turn entails that g is dictatorial on Z (note that it may not be dictatorial for profiles not restricted to Z).

Finally, consider the profile $\boldsymbol{x}^n = (z_{ba}, z_{ab}, z_{ab}, \dots, z_{ab})$. Clearly the optimal outcome is a, with $SC^* = SC(a) = 0.5 + \varepsilon + (n-1)\varepsilon = 0.5 + n\varepsilon$. On the other hand, the dictator mechanism returns $a_1^* = b$, and $SC(b) = (n-1)1 + 0.5 - \varepsilon = n - 0.5 - \varepsilon$. We get a cost approximation ratio of

$$\frac{SC(b)}{SC^*} = \frac{n - 0.5 + \varepsilon}{0.5 + n\varepsilon} = \frac{2n - 1 + 2\varepsilon}{1 + 2n\varepsilon} \geq 2n - 1 + 2\varepsilon - 2n\varepsilon = 2n - 1 - \varepsilon',$$

as required. □

If f is a duple (w.l.o.g. with range $\{a, b\} \subsetneq A$), then for the profile $\boldsymbol{x} = (c, c, \dots, c)$, $SC(f(\boldsymbol{x})) > 0 = (n-1)0 = (n-1)SC(c, \boldsymbol{x}) = (n-1)SC^*(\boldsymbol{x})$. Then $SC^*(\boldsymbol{x}) = SC(b, \boldsymbol{x}) = d(a, b)$, whereas $SC(f(\boldsymbol{x}), \boldsymbol{x}) = SC(a, \boldsymbol{x}) = (n-1)d(a, b)$. The proof makes use of the fact that the output of the mechanism is constrained to A, whereas agents' locations are not. Showing a similar bound for unconstrained facility location problems is much more challenging. We cover some such results in Section 5.3.3.

Randomized mechanisms Consider the *Random Dictator* mechanism g^{RD}, which selects each agent i with uniform probability and returns $a_i^* = \text{argmin}_{a \in A} d(a, x_i)$. The Random Dictator mechanism is clearly strategyproof, but how good is it? In contrast to Section 5.2 where the use of randomized voting mechanisms led to approximation ratio that depends on the number of alternatives, in facility location problems it guarantees a constant approximation ratio. Several

[3] That is, a mechanism that is strategyproof for any metric space $\langle \mathcal{X}, d \rangle$ and any $A \subseteq \mathcal{X}$.

variations of the problem where studied by Meir et al. [2012] and later by Feldman et al. [2016] and Anshelevich and Postl [2017].[4]

Theorem 5.9 [Meir et al., 2012]. *The Random Dictator mechanism has a cost approximation ratio of $3-\frac{2}{n}$ for any space $\mathcal{X}$, alternatives $A \subseteq \mathcal{X}$, and number of agents n. If the instance is unconstrained ($A=\mathcal{X}$) then the approximation ratio is $2-\frac{2}{n}$.*

Proof.

$$
\begin{aligned}
SC(g^{RD}(\mathbf{x})) &= \sum_{i\in N}\frac{1}{n}SC(a_i^*) = \frac{1}{n}\sum_{i\in N}\sum_{j\in N} d(x_j,a_i^*) \\
&\leq \frac{1}{n}\sum_{i\in N}\sum_{j\in N}(d(x_i,x_j)+d(x_j,a_j^*)) && \text{(triangle inequality)} \\
&\leq \frac{1}{n}\sum_{i\in N}\sum_{j\in N}(d(x_i,x_j)+d(x_j,c^*)) && \text{(by definition of } a_j^*\text{)} \\
&= \frac{1}{n}\sum_{i\in N}\sum_{j\neq i} d(x_i,x_j)+\sum_{j\in N} d(x_j,c^*) \\
&\leq \frac{1}{n}\sum_{i\in N}\sum_{j\neq i}(d(x_i,c^*)+d(c^*,x_j))+\sum_{j\in N} d(x_j,c^*) && \text{(triangle inequality)} \\
&= 2\sum_{i\in N} d(x_i,c^*)\sum_{j\neq i}\frac{1}{n}+\sum_{j\in N} d(x_j,c^*) = 2\sum_{i\in N} d(x_i,c^*)\left(1-\frac{1}{n}\right)+\sum_{j\in N} d(x_j,c^*) \\
&= \left(2\left(1-\frac{1}{n}\right)+1\right)\sum_{i\in N} d(x_i,c^*) = \left(3-\frac{2}{n}\right)SC(c^*) = \left(3-\frac{2}{n}\right)SC^*.
\end{aligned}
$$

To get the bound for unconstrained instances, note that in that case $d(x_j,a_j^*)=0$. □

Meir et al. [2012] provide variants of the random dictator mechanism that attain the same approximation ratios for social goals that attribute different importance (weights) to different agents.

Anshelevich and Postl [2017] introduce a parameter $\alpha \in [0,1]$ called *decisiveness*, which means that for each agent i, $d(a_i^*,x_i) \leq \alpha d(a,x_i)$ for any $a \neq a_i^*$. They prove an upper bound of $2+\alpha-\frac{2}{n}$ on the cost approximation ratio of the random dictator mechanism. Since $\alpha=0$ for unconstrained instanced and may get any value in $[0,1]$ for constrained instances, their result generalizes Theorem 5.9.

[4]Meir et al. [2012] phrase their theorems for binary classification problems. However, as they also show, binary classification of k samples is equivalent to facility location on the k-dimensional binary cube. Theorem 5.9 is proved by considering an arbitrary metric space (Theorem 4.4' there).

As it turns out, without further restrictions on the space $\mathcal{X}$, no strategyproof mechanism can do bette[redacted]an random dictator.

Theorem 5.1[redacted][Feldman et al., 2016, Meir et al., 2011]. *Any (randomized) strategyproof mechanism g has worst approximation ratio of at least $3 - \frac{2}{n}$.*

Meir et al. [2011] proved the theorem by an elaborate reduction that is using Gibbard's characterization of randomized strategyproof voting rules (note that such a reduction was also used for deriving bounds on the welfare approximation in Section 5.2). Feldman et al. [2016] later provided a much cleaner proof by identifying several implications of strategyproof location mechanisms. Notably, both proofs use constructions that require $n - 1$ dimensions, and thus it is not clear if a similar proof can be constructed for simpler spaces, or whether better mechanisms than the random dictator exist (note there is a gap between the upper and lower bound).

Table 5.1: Negative results (lower bounds) carry from left to right and from bottom to top. Positive results carry in the other direction.

	Unconstrained			Constrained		
Deterministic	Line	Circle	General	Line	Circle	General
Upper bound	1 [T. 5.11]	$n-1$	$n-1$ [T. 5.7]	3 [T. 5.12]	$2n-1$	$2n-1$ [T. 5.7]
Lower bound	1	$n-1$ [C. 5.15]	$n-1$	3 [T. 5.12]	$2n-1$ [Ex. 5.5(5)]	$2n-1$ [T. 5.8]
Randomized	Line	Circle	General	Line	Circle	General
Upper bound	1	$2-\frac{2}{n}$	$2-\frac{2}{n}$ [T. 5.9]	2 [T. 5.13]	$3-\frac{2}{n}$	$3-\frac{2}{n}$ [T. 5.9]
Lower b[redacted]nd	1	?	?	2 [T. 5.13]	2	$3-\frac{2}{n}$ [T .5.10]

5.3.2 LOCATION ON A LINE

When the p[redacted]ible points $\mathcal{X}$ are placed along a line, then each agent's preferences R_i induced by x_i are ne[redacted]sarily single-peaked w.r.t. the line (with x_i being the peak). Thus, the Median mechanism from Section 4.1 would be strategyproof. Interestingly enough, if the designer cares about minimizing social cost (sum of agents' negative utilities), then the Median mechanism is also optimal.

Theorem 5.1[redacted][Procaccia and Tennenholtz, 2009]. *For the unconstrained variant, when $A = \mathcal{X}$ are placed along the real line, $SC(g^{Median}(\mathbf{x})) = SC^*(\mathbf{x})$ for every profile $\mathbf{x}$.*

Note that we do not need to assume that $\mathcal{X}$ is finite, infinite, or equally spaced.

Proof. Note that $SC(c) = \sum_{i \in N} d(x_i, c)$. Suppose that agents are sorted and let $x^* = g^{Median}(\boldsymbol{x}) = x_{\lceil n/2 \rceil}$. Any point $x' > x^*$ increases the distance from all agents $1, 2, \ldots, \lceil n/2 \rceil$ by exactly $q \triangleq |x' - x^*|$. On the other hand, it decreases the distance to each other agent by at most q. Thus,

$$\begin{aligned}
SC(x') = \sum_{i \in N} d(x_i, x') &= \sum_{i \leq \lceil n/2 \rceil} d(x_i, x') + \sum_{i > \lceil n/2 \rceil} d(x_i, x') \\
&= \sum_{i \leq \lceil n/2 \rceil} (d(x_i, x^*) + q) + \sum_{i > \lceil n/2 \rceil} d(x_i, x') \\
&\geq \sum_{i \leq \lceil n/2 \rceil} (d(x_i, x^*) + q) + \sum_{i > \lceil n/2 \rceil} (d(x_i, x^*) - q) \\
&= \sum_{i \in N} d(x_i, x^*) + \lceil n/2 \rceil q - \lfloor n/2 \rfloor q \geq \sum_{i \in N} d(x_i, x^*) = SC(x^*).
\end{aligned}$$

Similarly, points below x^* increase the distance to all points above x^* and increase the social cost. □

Thus, for the standard utilitarian social welfare (or social cost) goal on the unconstrained line, the Median mechanism effectively solves everything. However, it is not the only strategyproof mechanism. While with general single-peaked preferences there are only few strategyproof mechanisms (see Section 4.1), the additional structure imposed by the distance metric allows for many more deterministic and randomized mechanisms. A complete characterization of all deterministic strategyproof mechanisms on a continuous line was provided by Schummer and Vohra [2004], and for a discrete line with equidistant vertices by Dokow et al. [2012].

We saw in Exercise 4.5(2) that the Median mechanism can be extended to *trees*. Proposition 5.11 extends to trees as well in a similar way.

Constrained locations on the line When the facility is restricted to specific locations $A \subsetneq \mathcal{X}$ on the line, the Median mechanism is no longer optimal, see Figure 5.2 for an example.

Theorem 5.12 [Feldman et al., 2016]. *When $\mathcal{X}$ are placed along the real line, the Median mechanism has a cost approximation ratio of 3 for any $A \subseteq \mathcal{X}$. No deterministic strategyproof mechanism does better.*

When randomized mechanisms are allowed, Feldman et al. [2016] propose the *Spike mechanism*, which is a lottery over order-statistic mechanisms. While the median gets the highest weight, the weights of all order-statistics are set in a sophisticated way.

Theorem 5.13 [Feldman et al., 2016]. *When $\mathcal{X}$ are placed along the real line, the Spike mechanism has a cost approximation ratio of 2 for any $A \subseteq \mathcal{X}$. No (randomized) strategyproof mechanism does better.*

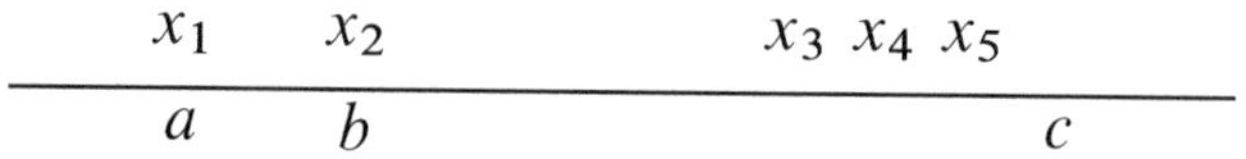

Figure 5.2: Consider the following instance on a line with $\boldsymbol{x} = (1, 2, 4.5, 5, 5.5)$, where the possible locations are constrained to $A = \{a = 1, b = 2, c = 6\}$. The optimal location for a facility is b, with $SC^* = SC(b) = 1 + 0 + 2.5 + 3 + 3.5 = 10$. However, the Median mechanism selects agent 3, with $SC(a_3^*) = SC(c) = 5 + 4 + 1.5 + 1 + 0.5 = 12 > SC^*$.

We note that both the deterministic and the randomized lower bounds hold even in very simple settings where $|A| = 2$. This was first proved in Meir et al. [2008, 2012] in the context of classification mechanisms.

Proof of the lower bounds. Consider $A = \{a = -1, b = 1\}$, and let $\varepsilon > 0$ be arbitrarily small. Consider the following three instances for two voters: $\boldsymbol{x} = (-1, \varepsilon)$, $\boldsymbol{x}' = (-1, 1)$, and $\boldsymbol{x}'' = (-\varepsilon, 1)$. Note that $SC(a, \boldsymbol{x}) = SC(b, \boldsymbol{x}'') = 1 + \varepsilon$, and $SC(a, \boldsymbol{x}'') = SC(b, \boldsymbol{x}) = 3 - \varepsilon$.

Assume toward a contradiction that there is a deterministic strategyproof mechanism g with approximation strictly better than $3 - 4\varepsilon < \frac{3-\varepsilon}{1+\varepsilon}$. Then $g(\boldsymbol{x}) = a$ and $g(\boldsymbol{x}'') = b$. Since g is strategyproof, $g(\boldsymbol{x}') = g(\boldsymbol{x}) = a$ as otherwise agent 2 has a manipulation by reporting $x_2' = \varepsilon$ in $\boldsymbol{x}$. Similarly, $g(\boldsymbol{x}') = g(\boldsymbol{x}'') = b$ as otherwise agent 1 has a manipulation by reporting $x_1' = -\varepsilon$ in $\boldsymbol{x}''$. Thus we get a contradiction.

Next, suppose that g is a randomized strategyproof mechanism with approximation ratio strictly better than $2 - 3\varepsilon$. Then $g(\boldsymbol{x})$ must select a with probability strictly greater than $\frac{1}{2}$, as otherwise $SC(g(\boldsymbol{x})) \geq \frac{1}{2}(1 + \varepsilon) + \frac{1}{2}(3 - \varepsilon) = 2 > (2 - 3\varepsilon)(1 + \varepsilon) = (2 - 3\varepsilon)SC^*(\boldsymbol{x})$. Similarly, $Pr(g(\boldsymbol{x}'') = b) > \frac{1}{2}$ and thus $Pr(g(\boldsymbol{x}'') = a) < \frac{1}{2}$. However, by the same strategyproofness argument as above, $Pr(g(\boldsymbol{x}) = a) = Pr(g(\boldsymbol{x}') = a) = Pr(g(\boldsymbol{x}'') = a)$, which is a contradiction. □

5.3.3 LOCATION ON A CIRCLE

A circle is an example of a metric space that seems only slightly more complicated than a line, but already captures much of the problems associated with designing a truthful voting mechanism. For an example where voters' preferences can be described as points on a circle, we can consider, for example, voting on an hour of the day in which we want to schedule some event or make some announcement (and agents each prefer this time to be as close as possible to their most convenient time).

Theorem 5.14 [Schummer and Vohra, 2004]. *The only deterministic strategyproof and onto facility location mechanism on the continuous circle is dictatorial.*

Schummer and Vohra also extend the theorem to any continuous graph that contains a cycle. When the circle is not continuous but is made of m equidistant nodes, the result depends on the size m: for sufficiently small m there are even anonymous—and in particular non-dictatorial—mechanisms (Exercise 5.5(6)), whereas for larger circles every strategyproof mechanism is "almost-dictatorial" in the sense that the facility can be placed at a distance of at most one edge from the dictator [Dokow et al., 2012].

It is easy to show that the cost approximation ratio of the dictatorial mechanism is at least $n-1$ even on unconstrained settings with only two candidates. We get the following theorem as an immediate corollary.

Corollary 5.15 **[Dokow et al., 2012, Schummer and Vohra, 2004]** *No deterministic strategyproof mechanism on a (continuous or large discrete circle) has a cost approximation ratio better than $n-1$, even in unconstrained instances.*

There is an open gap between the upper and lower bound for randomized mechanisms on unconstrained instances, and in particular it is still unknown whether the random dictator is the best possible randomized mechanism. See Table 5.1.

5.3.4 OTHER VARIATIONS

Many other variations of the facility location problem have been studied. We mention two directions that are more related to strategic voting. One is the search for mechanisms that are not just strategyproof, but group-strategyproof, and the other is an analysis of social goals beyond the sum of agents' costs. Both directions are explored in Alon et al. [2010], and we show one of their results below. Other directions such as mechanisms for placing multiple facilities [Escoffier et al., 2011], obnoxious facilities [Cheng et al., 2013] or mechanisms that allocate the cost of construction [Devanur et al., 2005] are farther away from the voting problems discussed in this book. See Cheng and Zhou [2015] for a survey.

Minimizing Egalitarian cost The *Egalitarian Cost* of $c \in A$ is the distance from the farthest agent, $EC(c, \mathbf{x}) = \max_{i \in N} d(c, x_i)$. As it turns out, the Median mechanism is not optimal for minimizing the Egalitarian cost.

Suppose that agents are ordered from left to right. The Left-Right-Median mechanism (g^{LRM}) returns the left-most peak (x_1) with probability $\frac{1}{4}$, the right-most peak (x_n) with probability $\frac{1}{4}$, and the median peak ($x_{\lceil n/2 \rceil}$) with the remaining probability of $\frac{1}{2}$.

Theorem 5.16 **[Alon et al., 2010].** *$EC(g^{LRM}(\mathbf{x})) \leq \frac{3}{2}EC^*(\mathbf{x})$ for every profile $\mathbf{x}$ on the real line. No randomized strategyproof mechanism can do better.*

A more nuanced fairness criterion is studied in a recent paper by Moulin [2017], where he designs a class of fair rules that are group-strategyproof for agents with convex single-peaked cost functions (Euclidean cost included). Informally, the mechanism starts from some arbitrary "benchmark" location, asks agents for their peaks, and selects a new location that minimizes the "damage" to the agent with the farthest peak.

5.4 JUDGMENT AGGREGATION

In judgment aggregation there is a set of "judges" who vote over a collection of interdependent issues. For general background see List and Pettit [2002], List and Puppe [2009], as well as a recent book chapter that reviews judgment aggregation in the broader context of social choice [Endriss, 2016].

In this section we focus on judgment aggregation where the judges are strategic, and more specifically on aggregation mechanisms that aim to preclude strategic or manipulative behavior.

5.4.1 FORMAL FRAMEWORK

Our notation roughly follows Lang and Slavkovik [2013] with some simplifications.

Let $\mathcal{B}_k$ be the set of all propositional formulas built from $k \geq 2$ propositional variables $V = \{v_1, \ldots, v_k\}$, using the usual connectives $\neg, \wedge, \vee, \rightarrow, \leftrightarrow$, and the constants $1, 0$ (corresponding to "true" and "false").

An *agenda* is a pair $\langle V, \Gamma \rangle$, where $\Gamma \subseteq B_k$. Each $v \in V$ is called an *issue*, and each $\gamma \in \Gamma$ is called a *constraint*.[5]

Given an agenda V, the *position* of every agent (or *judge*) assigns a truth value to every issue in V. Formally, a position is a subset $J \subseteq V$. We also abuse notation by writing J as a function $J : V \rightarrow \{0, 1\}$, where $J(v_i) = 1 \iff v_i \in J$. We say that J is *consistent* (with respect to Γ) if the assignment $J(V)$ does not violate any constraint in Γ.

Example 5.17 We can think of jury in a drunk driving case with three issues: v_1 corresponds to the whether the defendant was indeed driving, v_2 corresponds to whether the defendant was drunk, and v_3 corresponds to whether the defendant is guilty. The agenda has a single constraint $(v_3 \leftrightarrow v_1 \wedge v_2)$. The following table summarizes the positions of all three judges.[6]

	v_1 ("driving")	v_2 ("drunk")	v_3 ("guilty")
judge 1	1	1	1
judge 2	1	0	0
judge 3	0	1	0

We denote by $\mathcal{J} = \mathcal{J}(V, \Gamma) \subseteq 2^V$ the set of all *consistent* positions on agenda $\langle V, \Gamma \rangle$. We can think of agendas with a special structure imposed by the constraints Γ:

[5] Sometimes the issues themselves are also assumed to be logical formulae over the variables. We use a simplified presentation where all the relevant information on the agenda is in the constraints.

[6] This example, known as the *doctrinal paradox* or the *discursive dilemma* [List and Pettit, 2002], is often introduced in a non-strategic context. It demonstrates that when aggregating each issue independently (e.g., using the majority rule), the outcome—in this case $(1, 1, 0)$—may be inconsistent even if all individual opinions are.

- an agenda is *independent* if there are no constraints ($\Gamma = \emptyset$);
- the *conjunctive agenda* [Dietrich and List, 2007b] has a single constraint $\gamma = (v_k \leftrightarrow v_1 \wedge v_2 \wedge \cdots \wedge v_{k-1})$;
- two issues in an agenda are *connected* if there is a non-redundant constraint that involves both issues; and
- an agenda is *path-connected* if for any two issues there is a sequence of issues that contains both of them, such that every two consequent issues are connected.

Note that a conjunctive agenda is in particular path-connected. The agenda in Example 5.17 is conjunctive.

A *judgment aggregation rule* for an agenda $\langle V, \Gamma \rangle$ and judges N is a non-constant function $g : \mathcal{J}(V, \Gamma)^n \to \mathcal{J}(V, \Gamma)$. That is, it is a mapping from a (consistent) position profile $\boldsymbol{J} = (J_1, \ldots, J_n)$ to an outcome $g(\boldsymbol{J})$, which is also a consistent position in $\mathcal{J}(V, \Gamma)$.[7]

We define the following properties that an aggregation rule g may satisfy:

- g is *monotone* at issue $v_j \in V$ if changing the position of a voter $J_i(v_j)$ can only change the outcome $g(\boldsymbol{J})_j$ in the same direction; and
- g is *independent* at issue $v_j \in V$ if $g(\boldsymbol{L})_j$ depends only on $(J_i(v_j))_{i \in N}$.

Judgment aggregation vs. voting and facility location To see why judgment aggregation generalizes voting, consider a set of m alternatives A. We define an agenda where V contains $k = \binom{m}{2}$ issues, and each issue v_{ab} corresponds to a pair of distinct alternatives $a, b \in A$. For convenience we also denote $v_{ba} \triangleq \neg v_{ab}$. More specifically, $v_{ab} = 1$ means that $a \succ b$, whereas $v_{ab} = 0$ (or, equivalently, $v_{ba} = 1$) means that $b \succ a$. The constraint set Γ contains all *transitivity constraints*, that is, $\Gamma = \{(v_{ab} \wedge v_{bc} \to v_{ac})\}_{a,b,c \in A}$.

We thus have a one-to-one mapping between transitive preferences $\mathcal{L}(A)$ and consistent positions $\mathcal{J}(V, \Gamma)$. The outcome of an aggregation rule is a point $g(\boldsymbol{J}) \in \mathcal{L}(A)$, that is, an order rather than a single winner. We can take the winner to be $\text{top}(g(\boldsymbol{J}))$. Note that the agenda we obtained (the *transitive agenda*) is path-connected.

To see why judgment aggregation (and thus voting) is a special case of unconstrained facility location, note that given an agenda $\langle V, \Gamma \rangle$ over k variables we can define $\mathcal{X} \subseteq \{0, 1\}^k$ such that $\mathcal{X} = \{(J(v_1), \ldots, J(v_k)) : J \in \mathcal{J}(V, \Gamma)\}$ and $A = \mathcal{X}$. Note that any position J can be interpreted as a binary vector. The natural distance metric is the Hamming distance on the binary cube, which means that the distance between to positions J, J' is the number of issues on which they disagree.

[7]There are also versions where either the profile or the outcome may violate consistency [Endriss, 2016], or may be further restricted.

Therefore, any non-constant facility location mechanism for (subsets of) the binary cube, is also a judgment aggregation rule, and any judgment aggregation rule is a voting rule.[8] There are many interesting connections between axiomatic and other properties of such rules and mechanisms (see, for example, Dietrich and List [2007a], Dokow et al. [2012], Endriss [2016]) but these are beyond the scope of this book. Certain agendas give rise to specific aggregation rules that make sense. For example, under a conjunctive agenda, we can think of the *premise-based* aggregation rule g^{PB}, which for every $j \leq k - 1$ sets v_j to be the majority of $(J_i(v_j))_{i \in N}$, and then completes v_k in the unique consistent way. In contrast to facility location, there is typically no optimization goal, but rather we want aggregators that have certain axiomatic properties like IIA (that is, $g(x_1, \ldots, x_n)_j = f_j(x_{1j}, \ldots, x_{nj})$ for all $j \leq k$), unanimity ($g(x, \ldots, x) = x$) and so on.

5.4.2 INCENTIVES AND MANIPULATION

The main difference between judgment aggregation and the previous models we considered lies in the interpretation of preferences and incentives. In facility location problems we had a clear notion of (negative) utility, derived from the distance between the location of the agent and the facility. In voting, each voter had a well defined (ordinal or cardinal) order over alternatives.

Since there is no single notion of utility in judgment aggregation, manipulation is also not uniquely defined [Dokow and Falik, 2012]. If we follow the facility location approach with the Hamming distance as explained above, we get that agents are "utilitarian" and care about the total number of issues that match their opinion. Alternatively, we can think of an "optimistic" manipulation, where the agent succeeds in changing the outcome on one particular issue v_j to $J_i(v_j)$, regardless of how other issues are affected; "pessimistic," where the outcome must be at least as good for i on every issue v_j, or other manipulative goals. This is important, as different notion of utilities induce different sets of possible manipulations, and thus different definitions of a strategyproof aggregation mechanism.

Manipulability Consider a conjunctive agenda $< V, \Gamma >$ with k variables, and suppose that agents only care about the conclusion v_k (although their position is defined for all issues).

Consider Example 5.17. Note that if the premise-based aggregation rule is used and judges are truthful, the outcome will be $(1, 1, 1)$. However, if judge 2 only cares about the outcome v_3, she has the following manipulation: reporting $J'_2 = (0, 0, 0)$ (which is also a consistent position), will result in the outcome $(0, 1, 0)$, which judge 2 strictly prefers.

Dietrich and List [2007b] avoid the problem of defining explicit incentives and instead define *manipulability* directly, as the possibility of a judge to change the outcome of any issue to a more desired one (regardless of the effect on other issues).

Definition 5.18 (Manipulability). Given an agenda $\langle V, \Gamma \rangle$, an aggregation rule g, and a profile $\boldsymbol{J} \in \mathcal{J}^n$, g is *manipulable* at $\boldsymbol{J}$ by $i \in N$ on issue $v_j \in V$ if:

[8]More precisely, any judgment aggregation rule is a *social welfare function*, which in turn induces a unique voting rule.

1. $g(\boldsymbol{J})_j \neq J_i(v_j)$; and

2. there is $J_i' \in \mathcal{J}$ such that $g(\boldsymbol{J}_{-i}, J_i')_j = J_i(v_j)$.

Given a subset of issues $Y \subseteq V$, we say that g is *non-manipulable at* Y if g is not manipulable at any profile $\boldsymbol{J}$ by any $i \in N$ on any issue $v \in Y$.

Dietrich and List [2007b] provide a characterization of all non-manipulable aggregation rules on a given agenda. Intuitively, a rule is non-manipulable at Y if and only if it is monotone and independent on any issue $v \in Y$. Then, they prove the following impossibility result.

Theorem 5.19 [Dietrich and List, 2007b]. *For any path-connected agenda $\langle V, \Gamma \rangle$, an onto aggregation rule g is non-manipulable if and only if g is a dictatorship.*[9]

Back to preferences Without a well-defined *ordinal or cardinal* notion of preferences, we cannot analyze a judgment aggregation instance as a game (albeit any aggregation rule is a well-defined game form). Dietrich and List [2007b] suggest imposing *closeness-respecting* preferences, which can be intuitively thought of as the binary cube analogy of single-peaked preferences. In particular, preferences based on the Hamming distance are closeness-respecting (just like Euclidean distance is single-peaked in $\mathbb{R}^k$), but many other preferences can be closeness-respecting as well. Once we have preferences, we immediately get a definition of *strategyproofness*, which means that under no preference profile an agent can strictly gain by reporting a false position.

Then, Dietrich and List show that under the assumption that preferences are closeness-respecting, an aggregation rule is strategyproof if and only if it is non-manipulable (and thus must be either trivial or dictatorial). They also discuss the interesting relations of the above impossibility results with the G-S impossibility theorem for voting rules. The solutions they offer rely mainly on imposing certain structure on the agenda, and on applying partial aggregation. Note that neither one of these approaches has a natural translation to the general voting domain. In a follow-up work, Dokow and Falik [2012] characterize agendas and strategyproof aggregation rules according to more nuanced variations of preferences.

Different characterizations that rely on the abstract topology of the preferences rather than on the logical agenda are thoroughly studied in Nehring and Puppe [2010], and their connection to voting is more explicit. In fact, we presented some of these results in Section 4.1.2 when discussing median spaces. The complexity of manipulation in judgment aggregation (which corresponds to our Section 4.2 on complexity of voting manipulation) is studied in Endriss et al. [2012].

[9]The additional requirements of *universal domain*, *collective rationality* and *responsiveness* [Dietrich and List, 2007b] are satisfied by any onto aggregation rule g in our simplified model, and we thus avoid defining them explicitly.

5.5 EXERCISES

1. Compute the winner, the VCG payments, and voters' utilities for the following instance with four alternatives $\{a, b, c, d\}$ and three voters. Assume lexicographic tie-breaking. $U_1 = (3, 6, 4, 1), U_2 = (4, 2, 8, 4), U_3 = (6, 7, 2, 3)$.

2. Let g be a randomized voting rule. Let g' be an anonymous voting rule that first selects permutation π of N uniformly at random, and returns $f'(\boldsymbol{L}) = f(\pi(\boldsymbol{L}))$.

   (a) Show that if g is strategyproof then so is g'.

   (b) Show that the worst case welfare approximation ratio of g' if at least as good as that of g.

   (c) Show that neutrality also only improves the approximation ratio.

3. Suppose you are trying to approximate the social welfare of Range voting. What is the approximation ratio of the Random Dictator mechanism? What is the approximation ratio of the (non-strategyproof) Borda rule? Assume utilities are normalized as in Section 5.2.

4. In this question we show that the way we normalize utilities affects the approximation ratio. Suppose that agents' utilities are normalized so that $\sum_{a \in A} U_i(a) = 1$ (rather than setting the minimal and maximal values). Show that any deterministic strategyproof mechanism has an approximation ratio γ at most $\frac{1}{m(n-1)}$ (compare to Prop. 5.4—why is there a difference?).

5. Write an alternative proof for Theorem 5.8, but for the (restricted) case where $\mathcal{X}$ is a circle.

6. Find a non-dictatorial facility location mechanism for three agents on a discrete circle with seven vertices. (hint - think about the median)

7. Consider the following 5-issue agenda: $V = \{a, b, c, d, e\}$ with the constraints $\Gamma = \{d \leftrightarrow b \wedge c, e \leftrightarrow a \vee d\}$. What are all consistent positions $\mathcal{J}(V, \Gamma)$? Is the agenda path-connected? Give an example to a manipulation of issue e.

PART II

Voting Equilibrium Models

CHAPTER 6

Simultaneous Voting Games

Once we accept that voters are going to behave strategically, and think of voting rules (with preferences) as games, we can analyze them with game theoretic tools like any other game. In this chapter and the next ones, we will review some prominent attempts to "solve" voting games using familiar or novel equilibrium concepts. As these voting models are very different from one another in their assumptions and conclusions, it is hard to tell, a priori, if a given model is useful in some context or at all. Therefore we included in Section 6.1 criteria by which the various models can be evaluated. The rest of the chapter considers the normal form game induced by a voting rule f and a given population (a preference profile, or a distribution over profiles), and the various notions of equilibrium that may apply to such a game.

Section 6.2 looks at *implementation theory*, which aims at obtaining the truthful outcome of a voting rule in equilibrium when dominant strategies are not available. We then forgo the attempts to design a mechanism and analyze the expected outcome of common voting rules: Section 6.3 looks at the effect of slight modifications of voters' preferences to include the cost of voting or strategizing. Section 6.4 introduces the celebrated "calculus of voting" model where voters maximize their expected utility given a prior distribution of the population. In Section 6.5 we briefly overview several other approaches to define and analyze voting equilibrium.

Voting games that further diverge from the simultaneous-vote normal form description will be considered in the next two chapters.

6.1 DESIDERATA FOR VOTING MODELS

Meir et al. [2014] suggest several criteria by which we can evaluate various models and solution concepts. We bring here these desiderata and show how well some of them apply to the solution concept of Nash equilibrium, as an example. We let the reader judge how well the other models and solution concepts in the following chapters meet the criteria.

We will not be too picky about what is considered a "voting model" and whether it is described in terms of individual or collective behavior. The key feature of a model is that given a *preference profile*, it can be mapped to one or more *voting profiles*. We classify desirable criteria to the following classes: theoretic (mainly game-theoretic), behavioral, and scientific.

THEORETIC CRITERIA

Rationality. The model should assume that voters are behaving in a rational way in the game-theoretic sense, that is, trying to maximize their own utility based on what they know and/or believe.

Equilibrium. The model predicts outcomes that are in equilibrium, for example, a refinement of Nash equilibrium, or of another popular solution concept from the game theory literature. It is more appealing if equilibrium can be naturally computed or even reached by some natural dynamic (similar to the convergence of best reply dynamics to a pure Nash equilibrium in congestion games [Rosenthal, 1973]).

Discriminative/predictive power. The model predicts a small but non-empty set of possible outcomes. More specifically, it predicts a small set of possible winners, as there may be many voting profiles that have the same winner.

Broad scope. The model applies (or can be easily adapted) to various scenarios such as simultaneous, sequential or iterative voting, and to the use of different voting rules.

Sanity checks. Predicted outcomes should not include outcomes that are obviously unreasonable or absurd.

Nash equilibrium meets the criteria of Rationality and Equilibrium by its definition, and is well defined for any game (thus has a broad scope). On the other hand, we saw that for most common voting rules it has very weak discriminatory power and fails basic sanity checks, as almost any outcome is a Nash equilibrium.

BEHAVIORAL CRITERIA

By behavioral criteria, we mean what implicit or explicit assumptions the model makes on the behavior of voters in the game.

Voters' knowledge. Voters' behavior in the model should not be based on information that they are unlikely to have, or that is hard to obtain.

Voters' capabilities. The decision of the voter should not rely on complex computations, non-trivial probabilistic reasoning, etc.

We would like the behavioral assumptions of the model, whether implicit or explicit, to be supported by (or at least not to directly contradict) studies in human decision making.

Playing a Nash equilibrium requires voters to know the actions of all other voters and compute their best reply, which is simple enough. However, *finding* a Nash equilibrium in the general case requires knowing the exact preferences of all voters, or a distribution thereof, which is less likely.

SCIENTIFIC CRITERIA

Like any scientific theory that merits to explain some phenomena in the real world, models of strategic voting are subject to the following criteria.

Robustness. Except in a few knife-edge cases, we would not expect every small perturbation to change the identity of the winner. In particular, we expect the model to give similar predictions even if some voters do not exactly follow their prescribed behavior, if we slightly modify the available information, if we change order of players, and so on.

Few parameters. If the model has parameters, we would like it to have as few as possible, and we would like each of them to be meaningful (e.g., voters' memory, or risk aversion level). A related requirement is that the model will be *easy to fit*, given empirical data on voting behavior.

Reproduction. When simulating the model on generated or real preferences, we would like it to reproduce common phenomena such as *Duverger's law*.[1] We list more such phenomena with references to empirical and experimental literature in Section 9.1.

Prediction. The hardest test for a model is to try to predict the behavior of human voters based on their real preferences. By comparing the predicted and real votes (or even just outcomes), we can measure the accuracy of the model.

Nash equilibrium has no parameters, but may be quite sensitive to small variations of the game. As it has no discriminatory power, reproduction and prediction of human behavior are meaningless for most voting rules.

Lastly, some voting models explain how strategic behavior is *better for society*. For example, equilibrium outcomes in Plurality with a particular voter behavior may have a better Borda score, or coincide more with choosing a Condorcet winner. Similarly, there are cases where truthful voting results in a *paradox*—a winner that is clearly bad for society [Xia et al., 2011]. A model of strategic voting may show how voters overcome, or are still prone to, similar paradoxes. Although this is not exactly a criterion for a good model (as real strategic behavior may or may not increase welfare), we are interested in the conditions under which the theory predicts an increased welfare, as these may be useful for design purposes.

6.2 IMPLEMENTATION

Consider any (non-resolute) SCC F, in other words, a function that maps strict preference profiles to subsets of outcomes. In what follows, $F(\boldsymbol{L}) \subseteq A$ can be thought of as some set of socially desirable alternatives under preference profile $\boldsymbol{L}$. In this section we will sometime allow SCCs that return the empty set for certain profiles (and will mark them with G rather than

[1]Duverger's Law [Duverger, 1963, Riker, 1982] asserts that in Plurality voting, there will be two leading candidates getting almost all votes, whereas all other candidates will get a negligible amount of votes.

F). As usual, lower case f marks a resolute SCC. Some examples of SCCs we might want to implement are:

- F^{PAR}: All Pareto optimal alternatives in $\boldsymbol{L}$;
- f^{BL}: The (unique) Borda winner in $\boldsymbol{L}$, with lexicographic tie-breaking;
- F^{B}: All Borda winners of $\boldsymbol{L}$ (before tie-breaking);
- G^{CON}: All Condorcet winners of $\boldsymbol{L}$ (which may be empty);
- F^{NCL}: All alternatives that are not Condorcet losers;

and so on.

Intuitively, we can think of any implementation result as a triplet that consists of: (i) a mechanism g; (ii) a *target* SCC F; and (iii) some reasonable solution concept SOL, which specifies how voters may behave in any profile $\boldsymbol{L}$. "Reasonable" in that sense typically means a common game theoretic solution concept such as Nash or strong equilibrium. Note that we only consider strict preferences when discussing implementation. The mechanism $g : \mathcal{A} \to A$ may be any game form and is not necessarily a valid voting rule itself. In particular, the set of voter i's actions A_i may be arbitrary.

Denote all outcomes of g under the chosen solution concept SOL when voters have preferences $\boldsymbol{L}$ by $SOL_g(\boldsymbol{L}) \subseteq A$. A *positive result* means that $SOL_g(\boldsymbol{L}) = F(\boldsymbol{L})$ for any $\boldsymbol{L} \in \mathcal{L}(A)^n$. In contrast, a *negative result* is typically of the form "there is no mechanism g such that $SOL_g(\boldsymbol{L}) = F(\boldsymbol{L})$ for all $\boldsymbol{L}$."

A most natural question is *whether a voting rule implements itself* under some behavior. For example, if we take the Borda rule f^{BL}, we may look for a solution concept SOL such that $SOL_{f^{BL}}(\boldsymbol{L}) = \{f^{BL}(\boldsymbol{L})\}$ for all $\boldsymbol{L} \in \mathcal{L}(A)^n$. We can ask a similar question on any other voting rule f.

Implementation theory is a major topic in the economic literature. In fact, almost all of the results in this book could be framed as positive or negative implementation results. For example, the G-S theorem states that only dictatorial rules and duples implement themselves in truthful dominant strategies.[2] However, there is a big conceptual difference between truthful voting mechanisms and those that must rely on some sophisticated behavior, and hence the separation. The results in Part I of the book deal with truthful dominant strategies; this section deals with what targets (such as the examples above) can be implemented with common equilibrium concepts, and the rest of Part II looks at particular voting rules and mechanisms (Plurality, sequential voting, etc.) and asks what are the properties of their outcome, or what do they implement?. For an in-depth coverage of classic results in implementation theory, see Palfrey [2002].

[2]Some implementation results may add a fourth component, namely the set of allowed profiles. Thus for example, the *Median mechanism* (g^{median}) implements the *Condorcet winner* (G^{CON}) in *dominant strategies* for *single-peaked profiles on a line* (see Theorem 4.2).

Dominant strategy implementation Before diving into solution concepts that assume some sophistication, one may ask how much mileage we can get from dominant strategies. That is, for a given voting rule f, is there a mechanism g_f such that for any profile $\boldsymbol{L} \in \mathcal{L}(A)^n$, players in the game $\langle g_f, \boldsymbol{L}\rangle$ have dominant strategies $\boldsymbol{a_L} \in \mathcal{A}$ for which $g_f(\boldsymbol{a_L}) = f(\boldsymbol{L})$?

The *revelation principle* determines that if such a mechanism g_f exists, then there is another mechanism g'_f that implements f in *truthful dominant strategies*. To see why, note that the mechanism g'_f could use g_f as a black box, and for every profile $\boldsymbol{L}$ return $g_f(\boldsymbol{a_L})$. However, this means that $g'_f \equiv f$, so f itself must be truthful! We conclude that only voting rules that are already strategyproof (see Section 3.2) can be implemented in dominant strategies, and thus this solution concept does not add any implementation power.

6.2.1 NASH IMPLEMENTATION

Projecting our definition above for Nash equilibrium as our solution concept, we get the following definition.

Definition 6.1 (NE implementation). A mechanism g implements SCC F in Nash equilibrium (NE), if $NE_g(\boldsymbol{L}) = F(\boldsymbol{L})$ for all $\boldsymbol{L} \in \mathcal{L}(A)^n$.

Note that in particular, this would require Nash equilibria of g to have the same discriminatory power as F.

Observation 6.2 A dictatorship f implements itself in NE.

Proof. Let i^* be the dictator, thus $f(\boldsymbol{L}) = \text{top}(L_{i^*})$ for all $\boldsymbol{L}$. Let $\boldsymbol{L} \in \mathcal{L}(A)^n$ be a preference profile, and $\boldsymbol{a} \in \mathcal{L}(A)^n$ be an action profile in f. We have that $f(\boldsymbol{a}) = \text{top}(a_{i^*})$. Also, $\boldsymbol{a} \in NE_f(\boldsymbol{L})$ if and only if $\text{top}(a_{i^*}) = \text{top}(L_{i^*})$, as otherwise i^* can strictly gain. Thus $NE_f(\boldsymbol{L}) = \{\text{top}(L_{i^*})\} = \{f(\boldsymbol{L})\}$ for all $\boldsymbol{L}$. □

Are there less trivial examples of such voting rules? The answer is quite negative, even if we allow the designer to use arbitrary mechanisms.

Theorem 6.3 [Maskin, 1999]. *No voting rule except dictatorships and duples can be implemented in NE by any mechanism. In particular, no manipulable voting rule implements itself in NE.*

Maskin further showed that if we want to implement (irresolute) SCCs rather than SCFs, then results are more positive. A trivial example is the SCC $F^{ALL}(\boldsymbol{L}) \triangleq A$, which can clearly be implemented (e.g., by Plurality with $n \geq 3$ voters).

A less trivial example is the SCC which returns all Pareto-optimal outcomes of $\boldsymbol{L}$. That is, $F^{PAR}(\boldsymbol{L}) \triangleq \{a \in A : \neg\exists b \in A, \forall i \in N, b \succ_i a\}$.

Recall that Maskin monotonicity (Definition 3.4) means that if $a \in F(\boldsymbol{L})$ and in $\boldsymbol{L}'$ all voters only move a up their ranking (possibly changing order between other candidates), then $a \in F(\boldsymbol{L}')$.

No veto power is a slightly stronger condition than unanimity, stating that if $n - 1$ voters rank $a \in A$ at the top then $a \in F(\boldsymbol{L})$. For example, F^{PAR} is Maskin monotone and has no veto power (Exercise 6.6(1)).

Theorem 6.4 [Maskin, 1999]. *Consider an SCC F with $n \geq 3$ voters. F can be implemented in NE if it is Maskin-monotone and has no veto power.*

On the other hand, a slight relaxation of these conditions results in non-implementable SCCs.

The main idea of Maskin's proof is to construct a mechanism where each voter reports *all of* $\boldsymbol{L}$, rather than just L_i.[3] Voters also report another signal that is used for coordination. If all voters report the same $\boldsymbol{L}$, then the mechanism returns $F(\boldsymbol{L})$. If exactly one voter reports something else, the mechanism uses the rest of the reported preferences to "punish" the deviator, which guarantees that reporting the truth is a Nash equilibrium. If there is some greater disagreement, the mechanism uses the coordination signal in a clever way to make sure that at least one voter gains by deviating, so there are no other Nash equilibria. Some of these ideas are demonstrated in the (considerably simpler) proof of Theorem 6.6.

6.2.2 STRONG IMPLEMENTATION

Let $SE_g(\boldsymbol{L}) \subseteq A$ be all candidates that win in some strong equilibrium of mechanism g in profile $\boldsymbol{L}$. Recall that an equilibrium is strong if there is no subset of voters that can all gain by deviating.

Formally, a mechanism g implements a mapping $G : \mathcal{L}(A)^n \to 2^A$ in SE, if $SE_g(\boldsymbol{L}) = G(\boldsymbol{L})$ for all $\boldsymbol{L}$.

Note that we do not require that G is a valid SCC, as it may return the empty subset. An example of such a mapping is G^{CON}, which returns the (possibly empty) set of Condorcet winners of profile $\boldsymbol{L}$.

SE implementation was studied by Sertel and Sanver [2004], which in particular showed the following.

Theorem 6.5 [Sertel and Sanver, 2004]. *The Plurality voting rule f^{PL} implements G^{CON} in SE, for all odd n.*

Proof. Suppose first that $\boldsymbol{L}$ has a Condorcet winner $c^* \in A$. We need to show that c^* wins in some SE, and that no other candidate wins in SE. Consider the profile where all voters vote for c^*. A subset of voters I can only change the outcome if $|I| > n/2$ voters deviate to some other

[3] This makes Maskin's mechanism completely impractical of course, as preferences are rarely public information, and when they are, there is no need for a vote.

candidate c'. However, since c^* is a Condorcet winner, at least $n/2$ voters strictly prefer c^* over c', and therefore at least some of the voters in I would strictly lose by changing the outcome to c'. Thus this is an SE.

On the other hand, consider any profile in which c wins and c is not a Condorcet winner. There is some c' such that $> n/2$ voters prefer c' over c. All these voters that are not currently voting to c' can deviate to make c' the winner. Thus this is not an SE. In particular, if there is no Condorcet winner in $\boldsymbol{L}$, then there are no SEs. □

Moulin and Peleg [1982] connected implementation theory with the theory of *cooperative games*. Further connections between SE implementation and cooperative games are shown byHolzman [1986, 1987].

6.2.3 IMPLEMENTATION IN UNDOMINATED STRATEGIES

Every method of implementation implicitly or explicitly assumes some behavior on behalf of the agents (see Section 6.1). As Börgers [1991] keenly observes, both NE and SE implementations require voters to know the preferences of other voters (Maskin's solution is extreme in that voters must also *report* others' preferences), which is often impractical. In contrast, to play an undominated strategy a voter needs not know anything about the actions or preferences of others. Formally, g implements F in *undominated strategies* if for all $\boldsymbol{L} \in \mathcal{L}(A)^n$

$$F(\boldsymbol{L}) = \{g(\boldsymbol{a}) : \forall i \in N, a_i \text{ is undominated at } \boldsymbol{L}\}.$$

Jackson [1992] shows that *any* SCC (under very weak assumptions on preferences) can be implemented in undominated strategies. This assertion may sound suspicious, and Jackson readily admits that the mechanisms required to obtain this result have a "questionable feature." Indeed, note that we did not explicitly define what the *input* of a mechanism is. For Jackson's theorem (and for many other implementation theorems, see references and discussion therein), the voters have an infinite set of allowed actions (e.g., report an unbounded positive integer), and the mechanism exploits this in a clever but quite impractical way to "eliminate" outcomes not in $F(\boldsymbol{L})$. In particular such constructions allow for actions that are dominated but not by any undominated action, thereby losing much of the common sense and appeal that undominated strategies have in the first place.

Börgers [1991] showed how to get a bit more mileage from this approach by explicitly implementing the SCC $F^{TOPS}(\boldsymbol{L}) \triangleq \{\text{top}(L_i) : i \in N\}$. Note that F^{TOPS} returns a subset of Pareto optimal alternatives in $\boldsymbol{L}$ (and thus has more discriminatory power than F^{PAR}), is unanimous, and non-dictatorial. His result provides some insight into common techniques used in implementation theory and is therefore brought here.

Theorem 6.6 **[Börgers, 1991].** *There is a finite mechanism that implements F^{TOPS} in undominated strategies.*

Proof. In the suggested mechanism, each voter reports two pieces of information. One is her preferences L_i, and the other is an integer $x_i \in [n]$. The mechanism first selects an "ad hoc dictator" $i^* = \left(\sum_{i \in N} x_i \mod n\right) + 1$. Then the mechanism returns $\text{top}(L_{i^*})$.

It is not hard to verify that for any voter whose preferences are L_i and for any number x, the action $a_i = (L_i, x)$ weakly dominates all actions $a'_i = (L'_i, x)$ for any $L'_i \neq L_i$. This is since a_i is strictly better when $i^* = i$ and does not matter otherwise. Within the set $\{(L_i, x)\}_{x \in [n]}$ no action weakly dominates any other action, so they are all undominated.

Thus in any profile of undominated strategies, all voters report their true preferences (and some arbitrary numbers), which means that the mechanism returns $\text{top}(L_i) \in F^{TOPS}(\boldsymbol{L})$ for some $i \in N$.

In the other direction, for any $c \in F^{TOPS}(\boldsymbol{L})$ there is some $j \in N$ such that $c = \text{top}(L_j)$. Then for the (undominated) action profile $a_1 = (L_1, j - 1)$, $a_i = (L_i, n)\ \forall i > 1$, we get that $\left(\sum_{i \in N} x_i \mod n\right) + 1 = j$ is the selected dictator and the mechanism returns c. □

One could argue that the described mechanism is just a clever way to run a random dictatorship without any explicit randomization.

Unfortunately, such weird constructions are necessary for positive results. Informally, F is *semi-strategyproof* if a voter can never gain by lying when the outcome is selected from $F(\boldsymbol{L})$ by an adversary to the manipulator. If mechanisms are restricted to have finite inputs, then the any implementable SCC is semi-strategyproof [Jackson, 1992]. In particular this result means that every implementable voting rule is (exactly) strategyproof, and thus either dictatorial or a duple, just as with voting rules that are truthful or NE implementable.

Finally, Carroll [2014] showed how to construct SCCs for extremely simple domains that still require an arbitrarily large (but finite!) sets of strategies to be implemented in undominated strategies.

6.2.4 OTHER NOTIONS OF IMPLEMENTATION

Some notions of implementation not discussed here use *undominated Nash equilibria* [Palfrey and Srivastava, 1991, Sjöström, 1994], *mixed Nash equilibria* [Jackson, 1992, Mezzetti and Renou, 2012], *mediated equilibrium* [Peleg and Procaccia, 2010], and *virtual implementation of lotteries* [Abreu and Sen, 1991]. Some implementation notions such as *subgame perfect equilibrium implementation* [Abreu and Sen, 1990, Moore and Repullo, 1988] assume we can restructure the game, for example, by dividing the voting process to several stages. Such sequential voting games are also discussed in Section 7.5.

Other authors considered different equilibrium concepts and showed how certain common voting rules can be implemented (sometimes by a simple mechanism or by the same rule) under these equilibria. Some examples are: Veto implements itself under *protective equilibrium* [Barberà and Dutta, 1986]; Plurality implements itself under *demanding equilibrium* [Merlin and Naeve, 1999]; and Plurality implements Maximin under *maximal stability scores* [Falik et al., 2012].

6.3 FALLBACK STRATEGIES

As we saw, Nash equilibrium (even pure) is not very informative in most common voting rules, as voters typically cannot change the outcome. Suppose that a voter who is non-pivotal prefers to vote according to some fallback option. This option could be, for example, to vote truthfully, or not to vote at all. In this section we see how a slight bias toward the fallback option can substantially alter the set of equilibria.

6.3.1 TRUTH BIAS/PARTIAL HONESTY

Formally, we can think of truth bias as a negligible cost attached to any vote that is non-truthful. Such a cost may refer to the mental burden of finding a strategy, or the moral burden of manipulation.

Hopefully, even such a small behavioral bias may help eliminating many undesired equilibria (e.g., when all voters vote for bottom candidates). This idea has been formalized independently under the names *truth bias* [Meir et al., 2010] and *partial honesty* [Dutta and Laslier, 2010], which are formally equivalent. A precursor of these ideas appeared already in Matsushima [2008], albeit not in the context of voting.

Definition 6.7 (Truth bias equilibrium). Let $\langle f, \boldsymbol{L} \rangle$ be a voting game where f is a standard voting rule, and let $\boldsymbol{U} \in \mathcal{U}_L(A)$ some cardinal utility profile that fits $\boldsymbol{L}$. Let $\varepsilon > 0$ such that $\varepsilon < |U_i(c) - U_i(c')|$ for all $i \in N$ and $c, c' \in A$. Define a game G where $u_i(\hat{\boldsymbol{L}}) = U_i(f(\hat{\boldsymbol{L}}))$ if $\hat{L}_i = L_i$ (i is truthful), and $u_i(\hat{\boldsymbol{L}}) = U_i(f(\hat{\boldsymbol{L}})) - \varepsilon$ otherwise.

The profile $\hat{\boldsymbol{L}} \in \mathcal{L}(A)^n$ is an *equilibrium under truth bias* (or *partially honest Nash equilibrium* of $\langle f, \boldsymbol{L} \rangle$ if it is a pure Nash equilibrium of G.

In general, truth bias may eliminate *all* Nash equilibria: Consider for example Plurality with lexicographic tie-breaking, where voters preferences are $c \succ_1 a \succ_1 b \succ_1 d$ and $d \succ_2 b \succ_2 a \succ_2 c$. One can check that none of the 16 action profiles is an equilibrium under truth bias.

Obraztsova et al. [2013] provide characterizations of Nash equilibria in Plurality with truth bias. In particular they show:

Theorem 6.8 [Obraztsova et al., 2013]. *Given a preference profile* $\boldsymbol{L}$, *and* $c = f^{PL}(\boldsymbol{L})$ *the truthful outcome of Plurality.*

1. *It can be tested efficiently if* $\boldsymbol{L}$ *is an equilibrium under truth bias.*
2. *It is NP-hard to decide if there is some action profile* $\boldsymbol{L}'$ *s.t.* $f(\boldsymbol{L}') \neq c$ *and* $\boldsymbol{L}'$ *is an equilibrium under truth bias.*

This result implies that the non-truthful equilibria under truth bias have no simple characterization.

Thompson et al. [2013] studied the effect of truth bias using simulations. They generated many preference profiles and computed *all* their Nash equilibria (for Plurality) with and without truth bias. First of all, they showed that instances where no PNE exist are uncommon, and especially rare when there is an even number of voters. Yet, the number of PNEs drops dramatically from millions (i.e., almost all action profiles for 5 candidates) to typically under 30. These PNEs are also concentrated around 1 or 2 candidates that win much more often than the others. In addition, they show that the outcome under truth-biased PNEs is more socially efficient according to various criteria (Figure 6.1).

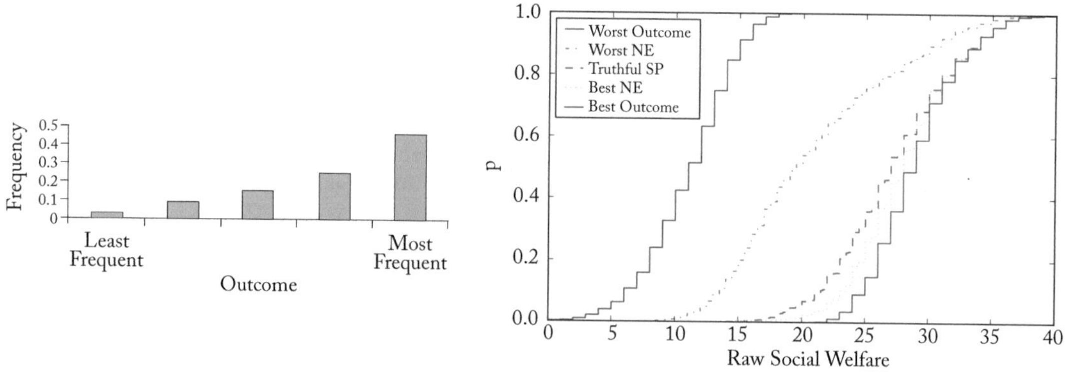

Figure 6.1: Figures based on Thompson et al. [2013]. On the left, we see the high decisiveness due to truth bias: about 45% of the truth-biased PNEs result in the same winning alternative (out of 5 alternatives). On the right, we can see that the average social welfare in a truth-biased PNE is very close to the social welfare in the best PNE.

We should note that as the number of voters grow, the probability that the *truthful* profile is a PNE tends to 1, both with and without truth bias.

In a recent follow-up work, Obraztsova et al. [2017] extend the truth bias model by assuming a different cost for any candidate, where candidates higher in the ranking have a lower cost according to some distance function. They provide some examples to situations where the distance function affects the outcome and show how it also affects the properties of equilibrium outcomes.

Implementation with partial honesty Truth bias/partial honesty was also studied in the context of implementation (see Section 6.2). It was shown that any SCC with $n \geq 3$ voters that satisfies the No Veto Power property can be implemented in partially honest Nash equilibrium [Dutta and Sen, 2012]. For additional implementation results with partial honesty see

Dutta and Sen [2012], Kartik et al. [2014] and Nunez and Laslier [2015] who focus on Approval with two voters.

6.3.2 LAZINESS AND THE PARADOX OF VOTING

Suppose that instead of experiencing ε cost for voting strategically, voters experience ε cost for voting at all, and may choose to abstain (or stay home). Since in most profiles all or almost all voters are non-pivotal, introducing such "lazy bias" means that almost no voter has any incentive to show up. This observation goes back to Condorcet and is widely known as the *paradox of voting* [Moulin, 1988, Owen and Grofman, 1984].[4]

The possibilities of abstention and voting costs are included in some of the models we discuss later in the chapter, which were partly motivated by the paradox.

Here we consider the effect of negligible cost ε on the Nash equilibria of the game $\langle f, \boldsymbol{L}\rangle$. Thus suppose each voter has an additional action of *abstaining* on top of A, which gives her a bonus of ε to her utility, regardless of the outcome. Also, ε is strictly smaller than the difference in utilities between any two possible outcomes, so voters will only choose to abstain when they cannot affect the outcome at all.

Under the Plurality rule, we see the paradox of voting in its full power. Note that the following results are independent of the preference profile $\boldsymbol{L}$.

Proposition 6.9 *Let $\boldsymbol{a}$ be a pure Nash equilibrium in Plurality voting with abstentions and lexicographic tie-breaking. Then at most one voter votes in $\boldsymbol{a}$ (all others abstain).*

Proof. Exercise 6.6(2). □

Desmedt and Elkind [2010] studied a variation of the problem, where ties are broken uniformly at random. Denote by $F^P(\boldsymbol{a})$ the set of candidates with maximal Plurality score, from which the winner is selected.

Theorem 6.10 [Desmedt and Elkind, 2010]. *Let $\boldsymbol{a}$ be a pure Nash equilibrium in Plurality voting with abstentions, and random tie-breaking.*

- *If $F^P(\boldsymbol{a}) = \{c\}$, then c gets a single vote whereas all other voters abstain.*
- *If $|F^P(\boldsymbol{a})| = k > 1$, then no voter abstains, and each candidate in $F^P(\boldsymbol{a})$ gets exactly $\frac{n}{k}$ votes.*

Proof. For the first part, clearly no voter i votes for $a_i \in A \setminus \{c\}$, since by abstaining she will get the same outcome and the ε bonus. Thus, since only c has active voters, if there is more than one voter with $a_i = c$, then this voter would also prefer to abstain.

[4]This is not to be confused with the *no-show paradox*, where a voter can manipulate the outcome by abstaining [Fishburn and Brams, 1983].

For the second part, let s be the Plurality score of the winner. If there is some voter i voting for $c \notin F^P(\boldsymbol{a})$, then i can gain ε by abstaining. If some voter j abstains, then j can gain strictly more than ε by actively voting for her favorite candidate in $F^P(\boldsymbol{a})$, making it the unique winner. □

Further, Desmedt and Elkind characterize all cardinal utility profiles for which a given subset $W \subseteq A$ may win in some equilibrium but show that for a given utility profile $\boldsymbol{U}$ it is NP-hard to decide whether $NE_f(\boldsymbol{U}) \neq \emptyset$. Some corrections to the characterization and additional results on variations of the problem appear in the follow-up papers [Elkind et al., 2015b, 2016].

Lazy bias in randomized voting rules The paradox of voting relies to a large extent on the determinism of the voting rule. It is obvious for example, that in a random dictatorship, every voter has a strict incentive to participate if the cost of voting is sufficiently low. In Brandl et al. [2015], the authors explore the possibilities and limitations of randomized voting rules that aim to satisfy various notions of participation and efficiency.

In particular, they focus on voting rules based on pairwise majority (such as Copland and Maximin), and show for example that no such rule is both unanimous and incentivize participation (voting truthfully weakly dominates abstention). On the other hand, they construct voting rules where participation strongly dominates abstention and hold weaker efficiency requirements.

6.4 THE "CALCULUS OF VOTING"

To demonstrate the ways various models deal with voters' beliefs and strategic decisions, we will introduce two running examples with specific populations. These examples will also be used in the later chapters.

Example 6.11 (Running example) We consider a scenario with four candidates, and a population of voters whose distribution is as follows.

Type	1	2	3	4	5
Preference order	*abcd*	*adbc*	*bcda*	*cdab*	*dbca*
Fraction of the population	5%	5%	30%	25%	35%

In models where this matters, we will further assume that voters attribute to candidates cardinal utilities of $(1, 0.4, 0.2, 0)$, according to their own preferences.
If not specified otherwise, we will assume that there is a finite set of n voters whose types are distributed *exactly* according to the distribution above (e.g., if $n = 300$ there are 15 type 1 voters). In some models we will assume the actual voters are *sampled* from the distribution above.

Example 6.12 (Simplified running example) We will also consider a simplified version of Example 6.11 where candidate c and type 4 voters do not exist, and thus the type distribution is $(\frac{1}{15}, \frac{1}{15}, \frac{6}{15}, 0, \frac{7}{15})$.

We have seen that when voters know exactly how others are going to vote, they rarely have an incentive to strategize (or to vote at all). For example, in Example 6.11 with $n = 100$ voters, no voter can influence the outcome at all. Yet it is known that people often do vote strategically, or at least try to [Regenwetter, 2006]. One possible explanation for this discrepancy is *uncertainty*: Since voters do not know exactly the preferences and actions of others, they know they *might* be pivotal, and hence some actions may be better than others in expectation.

6.4.1 THE EXPECTED VALUE OF VOTING

The classic game-theoretic approach for uncertainty, or *games with partial information*, assumes that each player's *type* (preferences/utilities) is sampled from some distribution and that this distribution is common knowledge. Thus each player knows her own type, and some distribution on the other players' types. In equilibrium, each player is playing a mixed strategy contingent on her type, that is, a best reply to the (mixed) joined strategy of all other players.

The first mathematical models of voting as economic games with partial information are due to Riker and Ordeshook [1968], who sought to explain the paradox of voting (see Section 6.3.2). The underlying idea of the "calculus of voting" is that a rational strategic voter would aim to maximize her expected utility $P_a \cdot U_i(a) - c$, by considering: the *probability* P_a that a given vote would be pivotal in favor of some candidate a; the utility $U_i(a)$ she gains *if* candidate a wins; and the cost c of voting.

Specific theories in this spirit were later developed by Riker [1976] and Cox [1987, 1994], whose main goal was to study the conditions under which a Duverger Law phenomenon should be expected, at least with three candidates and the Plurality voting rule. Note that in this setting there is essentially just one strategic decision that voters can make: whether to "compromise" and vote for their second-best candidate $L_i(2)$ when their favorite candidate $L_i(1)$ is expected to get the lowest number of votes.

Enter Palfrey Palfrey [1988] analyzed a similar model in detail and showed that each voter will have a simple decision threshold, which depends on the *cardinal utility* she assigns to the intermediate candidate (assuming that the maximal and minimal utilities are 1 and 0, respectively), and on the probability of being pivotal in each possible tie. Voters' types are cardinal utility scales for all candidates: $\mathcal{D}$ is a distribution over $\mathcal{U}(A)$, and each voter U is sampled i.i.d. from $\mathcal{D}$. Thus pivot probabilities can be inferred from a multinomial distribution whose parameters are determined by the fraction of each type in the population according to $\mathcal{D}$.

To better understand the model, we will analyze in detail Example 6.12 with an i.i.d. sample of $n = 10$ voters and randomized tie-breaking under Plurality, from the perspective of voter i of type 1. Suppose first that all other nine voters are truthful.

The probability of an outcome where candidates scores are (s_a, s_b, s_d) is exactly $\binom{9}{s_a, s_b, s_d}(\frac{2}{15})^{s_a}(\frac{6}{15})^{s_b}(\frac{7}{15})^{s_d}$. We can list all states where the voter is pivotal and write down the utility of each action in each possible state. For example, the probability of the state $(5,4,0)$ is $\binom{9}{5,4,0}(\frac{2}{15})^5(\frac{6}{15})^4 = 0.00014$. In this state, a wins if i votes for a or d, or abstains, in which case i's utility (not including the cost of voting) is 1. If i votes for b, then a and b win with equal probability, in which case the utility for i is $\frac{1+0.4}{2} = 0.7$. Table 6.1 presents the probabilities and conditional utilities of each action in all possible states.

The horizontal lines separate the states in which i is pivotal between a, b, between a, d and between b, d. In all other states (e.g., $(7,1,1)$) the voter is not pivotal, so her vote does not affect expected utility. In this example, voting for d is clearly a bad idea (note that it is globally dominated by abstaining). Voting for b yields a slightly higher expected utility than truthfully voting for a, which justifies strategic voting. Note that there is a threshold for $U_i(b)$ above which the strategic vote pays off.

However, if all voters are strategic and make these considerations, then the probability of a voter voting to a is no longer $\frac{2}{15}$, which may affect the strategic decisions (i.e., the decision thresholds), which again affect the pivot probabilities and so on. An equilibrium is a collection of pivot probabilities and decision thresholds that *justify one another*. In the example above, the only equilibrium where both b and d get votes is where all type 1 and 3 voters vote for b, and all type 2 and 5 voters vote for d. In particular, no voter votes for a. The equilibrium expected utilities are as follows (all other states either occur w.p. 0, or a single voter is not pivotal in them).

Table 6.1: Pivotal states in Example 6.12 with 9 i.i.d. voters. The utilities are from the perspective of a type 1 voter.

Probability	scores	vote a	vote b	vote d	abstain
0.00014	$(5,4,0)$	1	0.7	1	1
0.00041	$(4,5,0)$	0.7	0.4	0.4	0.4
0.00238	$(4,4,1)$	1	0.4	0.7	0.7
0.00555	$(4,3,2)$	1	0.7	1	1
0.01665	$(3,4,2)$	0.7	0.4	0.4	0.4
0.06119	$(0,5,4)$	0.4	0.4	0.2	0.4
0.07139	$(0,4,5)$	0	0.2	0	0
0.10199	$(1,4,4)$	0.2	0.4	0	0.2
0.05828	$(2,4,3)$	0.4	0.4	0.2	0.4
0.06799	$(2,3,4)$	0	0.2	0	0
0.00088	$(4,0,5)$	0.5	0	0	0
0.00025	$(5,0,4)$	1	1	0.5	1
0.00378	$(4,1,4)$	1	0.5	0	0.5
0.02266	$(3,2,4)$	0.5	0	0	0
0.00648	$(4,2,3)$	1	1	0.2	1
0.02590	$(3,3,3)$	1	0.4	0	$\frac{1.4}{3}$
Expected utility		0.1364	0.1472	0.0395	0.1031

Probability	scores	vote a	vote b	vote d	abstain
0.226	$(0,5,4)$	$U_i(b)$	$U_i(b)$	$\frac{U_i(b)+U_i(d)}{2}$	$U_i(b)$
0.258	$(0,4,5)$	$U_i(d)$	$\frac{U_i(b)+U_i(d)}{2}$	$U_i(d)$	$U_i(d)$
Expected utility for type 1		0.0904	* 0.142	0.045	0.0904
Expected utility for type 2		0.1032	0.0516	* 0.1484	0.1032
Expected utility for type 3		0.3292	* 0.4066	0.2614	0.3292
Expected utility for type 5		0.3484	0.271	* 0.4162	0.3484

We can see that the optimal action for each type (marked with $*$) is the action we specified, so this is indeed an equilibrium. Further, we can also see that voting is better than abstaining even with a positive cost (e.g., with $c < 0.04$), which provides some explanation for the paradox of voting. However, for large n the gain from voting becomes negligible.

For a finite set of voters with arbitrary cardinal preferences, the calculation of the pivot probabilities becomes hairy when n grows, and computing an equilibrium may be difficult. Yet, in the limit when the number of voters is large (and U is fixed), the probability that the least-popular candidate will be tied for victory becomes negligible, and thus no voter can ever be pivotal for the least popular candidate. As a result, there is no equilibrium in which more than two candidates get a positive number of votes, in other words, Duverger's law should always be expected [Palfrey, 1988]. This last result was later debuted in a more careful examination by Fey [1997], see below. Exercise 6.6(3) lets the reader analyze the equilibria in another simple example.

Enter Myerson and Weber Myerson and Weber substantially generalized the "calculus of voting" model with the help of some simplifying assumptions [Myerson and Weber, 1993]. The crucial observation (also used by Palfrey [1988]) is that for most common voting rules, and in particular for PSRs, there is no need to guess the exact actions of all other voters. To assess the value of every action, it is sufficient to consider the probability that the voter is pivotal between every pair of candidates. In scoring-based voting rules, each rule can be defined by specifying a set of valid scores $V \subseteq \mathbb{R}^m_+$ as the actions available to each voter. Every action of a voter is a vector $v \in V$, meaning $v(x)$ points are added to the score of each candidate $x \in A$. For example, in Approval V contains all binary vectors, and in Plurality V contains vectors with a single "1" entry.[5]

Formally, suppose that for any $x, y \in A$, p_{xy} is the probability that x, y are tied and all other candidates have a strictly lower score. Myerson and Weber make a simplifying assumption here that if the voter assigns $v(y) > v(x)$ points to candidates y, x respectively, then the probability the winner changes from x to y is proportional to $p_{xy}(v(y) - v(x))$ (note that this is approximately true for large n). The expected *gain* (utility compared to abstaining) for the voter from voting v when her preferences are U:

$$\begin{aligned}\text{gain}(\boldsymbol{p}, U, v) &\triangleq \sum_{x,y \in A} [\text{probability outcome changes from } x \text{ to } y](U(y) - U(x)) \\ &= \sum_{x,y \in A} p_{xy}(v(y) - v(x))(U(y) - U(x)). \qquad (6.1)\end{aligned}$$

In Plurality this can be somewhat simplified. The gain from voting for $a \in A$ is:

$$\text{gain}(\boldsymbol{p}, U, a) = \sum_{x \in A} p_{xa}(U(a) - U(x)).$$

Equation (6.1) had been previously suggested by Merrill [1981], who observed that under many voting rules it captures a linear program whose solution obtains the voter's best reply and studied the conditions under which it is unique.

From here, Myerson and Weber adopt standard game-theoretic assumptions: $\mathcal{D}$ is a distribution over a finite number of voter types T, where $\mathcal{D}(t)$ is the fraction of type t voters.

A *voting profile* must specify what fraction of each type is using which strategy. Formally, it is a distribution $\boldsymbol{q} \in \Delta(T \times V)$ such that for every type $t \in T$, $\sum_{v \in V} q(t, v) = \mathcal{D}(t)$.

[5]This captures all PSRs, Approval, Range voting and so on.

Thus, an *equilibrium* in a given population (i.e., a distribution $\mathcal{D}$ over types) can be described by a voting profile $\boldsymbol{q}$, candidates' scores $\boldsymbol{s} \in \mathbb{R}^m_+$, and a probability vector $\boldsymbol{p} \in \Delta(A^2)$, such that:

- for every voter of type $t \in T$ and action $v \in V$, $q(t, v) > 0$ only if v maximizes $\text{gain}(\boldsymbol{p}, U_t, v)$;
- for every candidate $x \in A$, $s(x) = \sum_{t \in T} \sum_{v \in V} q(t, v)v(x)$; and
- $p_{xy} \sim \frac{1}{k}$ if (x, y) is one of the pairs whose probability to be tied is maximal (k is the number of such pairs), and otherwise p_{xy} is (close to) 0.

The first bullet dictates that each voter is voting in a way that maximizes her expected gain; the second means that the score of each candidate equals its total number of votes from all types. The last bullet requires some clarification. Note that in either case, if we sample n voters from distribution $\mathcal{D}$, and they each vote according to the first bullet, then $\frac{1}{n} \sum_{i \leq n} v_i(x)$ is concentrated around $s(x)$. In the limit, the score x is exactly $s(x)$, so we only care about the pairs most likely to be tied (thus Myerson and Weber get almost for free the part that was most difficult in the analysis of Palfrey for finite n).

There are two cases determining which ties are most likely. Let $W \subseteq A$ be such that $s(w) = s^*$ for all $w \in W$ and $s(c) < s^*$ for all other candidates (the winner set). Case (a): $|W| > 1$, in which case the most likely ties are among all $k = \binom{|W|}{2}$ pairs from W; (b) $W = \{w\}$, in which case the most likely ties are between w and any of the k candidates whose score is maximal among $A \setminus W$. In Example 6.11 with a very large population, a voter in the MW model will only consider the possible tie between b and d, if she assumes other voters are truthful.

Thus w.l.o.g., for each $L \in \mathcal{L}(A)$, a symmetric profile $\boldsymbol{v}$ only needs to specify a single cutoff point v_L: any voter i such that U_i fits L (i.e., $L_i = L$) will vote for $\text{top}(L_i)$ if $U_i(2) < v_L$, and otherwise vote for her second-preferred candidate $L_i^{-1}(2)$.

A natural question is whether such an equilibrium exists.

Theorem 6.13 **[Myerson and Weber, 1993].** *For any scoring-based rule (including PSRs), and utility profile $\boldsymbol{U}$, there exists a voting equilibrium.*

In fact, for Plurality the existence of an equilibrium is trivial: Consider *any* two candidates $x, y \in A$, then there is an equilibrium where each voter votes for her more preferred candidate among x, y. A non-Duverger equilibrium where more than two candidates get votes is also possible but requires careful construction [Fey, 1997, Myerson and Weber, 1993].

See Myatt [2007] for a critique on the informational and behavioral assumptions underlying the calculus of voting. In particular, he shows that the construction of non-Duvergerian equilibria requires same-type voters to adopt distinct and seemingly unreasonable strategies. Myatt [2007] offers an alternative model where voters have improper prior over other voters' types, and where their beliefs might be correlated with their own preferences. Myatt restricts attention to equilibria that are symmetric (ignore voter's identity), have full support, and respond positively to the voter's information. He shows that there exists a unique equilibrium under these conditions.

6.4.2 EQUILIBRIUM STABILITY

Some of the criticism on calculus of voting models lies in the requirement (which is standard in game theory) that all voters know exactly the prior distribution and the equilibrium strategies of others. Although voters have no incentive to leave an equilibrium once it is reached, it is not clear how they are supposed to reach it in a game that is only played once without some means for coordination. Iterative voting (see next chapter) is one way around this, but here we present some solutions closer to the Bayesian models above.

Suppose that for some utility distribution $\mathcal{D}$ over $\mathcal{U}(A)$, the voting profile $\boldsymbol{q}^0$ is *not* an equilibrium. This means that if voters happen to vote $\boldsymbol{q}^0$ and scores $\boldsymbol{s}^0$ are published, then some voters have an incentive to change their vote. If all voters play their best reply to $\boldsymbol{s}^0$ (the strategy maximizing their gain), we get a new voting profile (partition) $\boldsymbol{q}^1 = h(\boldsymbol{q}^0)$. This profile may in turn not be an equilibrium either, which leads to $\boldsymbol{q}^2 = h(\boldsymbol{q}^1)$ and so on. This process may either converge to some voting profile $\boldsymbol{q}^* = h(\boldsymbol{q}^*)$ (which must be an equilibrium), or diverge.

Palfrey and Rosenthal [1990] classified equilibria according to this dynamic behavior in games where the partition $\boldsymbol{q}$ is a single number, including Plurality games with three candidates. We bring here a generalized but informal version of their classification:

- $\boldsymbol{q}^*$ is an *expectationally stable equilibrium* (ESE) if there is some convex set of voting profiles $I(\boldsymbol{q}^*)$ such that $\boldsymbol{q}^* \in I(\boldsymbol{q}^*)$ and for any $\boldsymbol{q} \in I(\boldsymbol{q}^*)$, $h(\boldsymbol{q})$ is *closer to* $\boldsymbol{q}^*$ than $\boldsymbol{q}$;

- $\boldsymbol{q}^*$ is a *globally expectationally stable equilibrium* (GESE) if the above holds with $I(\boldsymbol{q}^*)$ as the set of *all* voting profiles;

- $\boldsymbol{q}^*$ is an *expectationally unstable equilibrium* (EUE) if there is some convex set of voting profiles $I(\boldsymbol{q}^*)$ such that $\boldsymbol{q}^* \in I(\boldsymbol{q}^*)$ and for any $\boldsymbol{q} \in I(\boldsymbol{q}^*)$, $h(\boldsymbol{q})$ is *further from* $\boldsymbol{q}^*$ than $\boldsymbol{q}$; and

- $\boldsymbol{q}^*$ is a *globally expectationally unstable equilibrium* (GEUE) if the above holds with $I(\boldsymbol{q}^*)$ as the set of *all* voting profiles.

Thus, if the initial voting profile happens to be in $I(\boldsymbol{q}^*)$ of some ESE $\boldsymbol{q}^*$, then each "poll" will bring voters closer and closer to $\boldsymbol{q}^*$ until convergence. In contrast, even a slight deviation from an EUE will result in drifting away (possibly to a different equilibrium). For example, the Duvergerian equilibrium we found in Section 6.4.1 for Example 6.12 where a gets no votes is an ESE, since even in states near the equilibrium where a gets a small fraction of votes, still no voter has an incentive to vote for a. However it is not a GESE since a there is another ESE where no voter votes for b.

Fey [1997] shows that for Plurality voting with three candidates and sufficiently large n, Duverger equilibria are ESE, whereas all non-Duverger equilibria are EUE and thus inherently unstable.

6.4.3 SOCIAL NETWORKS

Clough [2007] suggests a variation of the classic calculus of voting models, where voters with single-peaked preferences are connected in a social network and gain all their information from their friends (i.e., they are playing the action that is best-in-expectation, assuming that the votes of their friends are representative of the entire population). The author shows via simulations that sometimes votes do converge to an equilibrium with more than two candidates getting votes.

In a follow-up paper, Tsang and Larson [2016] ran additional simulations on various classes of networks and measured the effect on voters' social welfare. They showed that the amount of strategizing in Plurality increases with the edge density of the graph until it reaches saturation. Moreover, increased strategizing leads to a more substantial concentration on two candidates (Duverger's law). Interestingly, in *homophilic* graphs (i.e., when social connections are based on proximity of preferences), honest voting results in a higher social welfare than strategic voting, whereas in graphs with independent social connections we see the opposite—the (best) equilibrium outcome results in higher social welfare than truthful voting.

In an independent line of work, Sina et al. [2015] considered voting over a social network but focused on how to modify the network in favor of a particular candidate.

6.4.4 QUANTAL RESPONSE EQUILIBRIUM

Quantal response of a decision maker is a "soft" randomized version of utility maximization where the probability to play each action is increasing in its (expected) utility [McKelvey and Palfrey, 1995].[6]

A quantal response equilibrium (QRE) in a game $\langle f, \boldsymbol{U} \rangle$ for a given noise distribution Δ is a profile of mixed strategies $\boldsymbol{\sigma}$ such that any strategy σ_i is the quantal response (according to Δ) to $\boldsymbol{\sigma}_{-i}$.

McKelvey and Patty [2006] applied QRE analysis to Plurality voting. Their starting point is a model similar to that of Palfrey [1988] (i.e., the finite version of the calculus of voting), but where voters each play quantal response instead of optimal response. Thus even voters of the same type may vote differently, with more voters (in expectation) choosing the better actions. They prove that a *voter equilibrium*, that is, a QRE, exists for any utility profile $\boldsymbol{U}$.

[6] Another interpretation is that the decision maker acts deterministically but observes the utilities of each action with random noise.

6.5 OTHER EQUILIBRIUM MODELS

6.5.1 MINIMAX REGRET

The *minimax regret* (MMR) criterion due to Savage [1951] stipulates that in face of uncertainty, decision makers aim to minimize their maximal regret from choosing the "wrong" action in retrospect rather than to maximize their expected utility. This decision criterion is particularly attractive in situations where probabilistic estimations are difficult. Indeed, Ferejohn and Fiorina [1974] argued that voting is one such case and developed a theory of strategic voting that minimizes the worst-case regret of the voter.

The theory that voters vote so as to minimize their regret was at some point the main competitive theory to the "calculus of voting" as a formalization of Downs' ideas.

In the MMR model, each voter has cardinal utility U_i over candidates, and in addition bears a fixed cost $c < \frac{1}{2}$ for casting a ballot. In each state of the world, it is possible to say exactly what is the utility of each action, according to whether the voter is pivotal and among which candidates.

Consider Table 6.1 presenting the pivotal states in Example 6.12. If we add a cost c for voting and ignore the pivot probabilities we get Table 6.2. We only consider states that are possible with nine voters. Having an arbitrary number of voters would add a few more states.

Table 6.2: Pivotal states in Example 6.12 and utilities for a type 1 voter. t is the score of the winner(s) and $s < t$. The best action in each state is marked with (*).

scores	vote a	vote b	vote d	abstain
$(t, t-1, s)$	$1-c$	$0.7-c$	$1-c$	(*) 1
$(t-1, t, s)$	(*) $0.7-c$	$0.4-c$	$0.4-c$	0.4
(t, t, s)	(*) $1-c$	$0.4-c$	$0.7-c$	0.7
$(s, t, t-1)$	$0.4-c$	$0.4-c$	$0.2-c$	(*) 0.4
$(s, t-1, t)$	$0-c$	(*) $0.2-c$	$0-c$	0
(s, t, t)	$0.2-c$	(*) $0.4-c$	$0-c$	0.2
$(t-1, s, t)$	(*) $0.5-c$	$0-c$	$0-c$	0
$(t, s, t-1)$	$1-c$	$1-c$	$0.5-c$	(*) 1
(t, s, t)	(*) $1-c$	$0.5-c$	$0-c$	0.5
$(3,3,3)$	(*) $1-c$	$0.4-c$	$0-c$	0.466
all other	$X-c$	$X-c$	$X-c$	(*) X

Note that without assigning explicit probabilities to the different states, it is not possible to calculate expected utilities. Instead, the minimax regret criterion suggests that the decision maker compares each action to the *best possible action in retrospect* to derive its maximal regret. In every row in the table above, the best action (column) is marked with *, assuming for the purpose of this example that $c < 0.2$. Table 6.3 shows the regret for every action in every state, which is the *difference* between the utilities of the optimal action and chosen action. Note that the regret of the optimal action is always 0.

The maximal regret of each action $a_i \in A_i$ is computed as

$$\text{MaxRegret}(a_i, U_i) \triangleq \max_{\boldsymbol{a}_{-i} \in \boldsymbol{A}_{-i}} \left(\max_{a_i^* \in A_i} U_i(f(\boldsymbol{a}_{-i}, a_i^*)) - U_i(f(\boldsymbol{a}_{-i}, a_i)) \right),$$

and voter i is assumed to select an action minimizing $\text{MaxRegret}(a_i, U_i)$. One can check that the conclusion of Table 6.3 would remain the same for a large number of voters.

Table 6.3: Pivotal states in Example 6.12, and regret for a type 1 voter.

scores	vote a	vote b	vote d	abstain
$(t, t-1, s)$	c	$c + 0.3$	c	(*) 0
$(t-1, t, s)$	(*) 0	0.3	0.3	$0.3 - c$
(t, t, s)	(*) 0	0.6	0.3	$0.3 - c$
$(s, t, t-1)$	c	c	$0.2 + c$	(*) 0
$(s, t-1, t)$	0.2	(*) 0	0.2	$0.2 - c$
(s, t, t)	0.2	(*) 0	0.4	$0.2 - c$
$(t-1, s, t)$	(*) 0	0.5	0.5	$0.5 - c$
$(t, s, t-1)$	c	c	$0.5 + c$	(*) 0
(t, s, t)	(*) 0	0.5	1	$0.5 - c$
$(3, 3, 3)$	(*) 0	0.6	1	$0.633 - c$
all other	c	c	c	(*) 0
max regret	0.2	0.6	1	$0.633 - c$

We can see that the action "vote a," in other words, to vote truthfully, minimizes the maximal regret of type 1 voters in this example.

Ferejohn and Fiorina [1974] analyze a general three-candidate scenario and derive a condition on c and on the utility of the intermediate candidate $v = U_i(L^{-1}(2))$, under which voting truthfully is optimal in terms of minimizing regret. In particular the condition holds if $c + v < 0.5$, in other words, it justifies truthful voting even with non-negligible cost c. Interestingly, the model predicts that a higher utility for the intermediate candidate v will not result in strategic compromise, but rather in a higher chance that the voter will abstain.

6.5.2 ROBUST EQUILIBRIUM

Even if the voting profile is known, we can introduce uncertainty in a finite set of voters as follows. For every voter $i \in N$, we do not count the vote of i with some small independent probability ε. This may be justified as a probability that the vote is "miscounted," or that the voter fails to vote [Messner and Polborn, 2005].

Let us focus on Plurality voting. For an action profile $\boldsymbol{a} \in A^n$, denote by $\boldsymbol{p}_\varepsilon(\boldsymbol{a}) \in \Delta(A^n)$ the distribution over profiles attained by eliminating each vote with i.i.d probability ε. Consider our Example 6.11 again, say with $n = 100$ voters *whose preferences follow exactly the given distribution*. Thus in the truthful profile $\boldsymbol{a}^0$ there are $s^0_a = 10$ votes for a, 30 votes for b, 25 for c, and 35 for d. We get pivot probabilities similar to those in Table 6.1, except that instead of multinomial distribution, each score s_x is sampled independently from a Binomial distribution $Bin(1-\varepsilon, s^0_x)$.

Note for example that for any $\varepsilon > 0$, voting for i's least preferred candidate b is strictly dominated, since with positive probability it is tied with some other candidate. Also, as in Section 6.4.1, the probability that a, b are tied is much higher than any other tie, and this is still true when $\varepsilon \to 0$. Thus a type 1 voter will be better off voting for b.

For a preference profile $\boldsymbol{L}$, a *robust equilibrium* (RE) is an action profile $\boldsymbol{a} \in A^n$ s.t. each a_i is a best reply to $\boldsymbol{p}_\varepsilon(\boldsymbol{a})$ for $\varepsilon \to 0$. This notion is closely related to *trembling hand equilibrium* in the game theory literature [Selten, 1975].

As with the voting equilibrium of the MW model above, for any two candidates $x, y \in A$ under Plurality, the profile where each voter votes for the more preferable candidate among x, y is a robust equilibrium. Messner and Polborn show that these are the *only* robust equilibria, thereby providing some additional theoretical justification to the Duverger Law.

A similar model of trembling-hand equilibrium was more recently analyzed in Obraztsova et al. [2016b]. The main difference is that with probability ε the voter submits a random ballot rather than failing to vote at all.

6.5.3 ITERATED REMOVAL OF DOMINATED STRATEGIES

Let us consider again games with complete information. Some actions may be *weakly dominated* and therefore highly unlikely to be played. If all players know other players will refrain from playing their weakly dominated strategies, we get a smaller game where only a subset of actions are present. In this new game it is possible that other strategies are weakly dominated, so we can remove those as well, and so on. Games where this process ends with a single action profile are known as *dominance solvable games* [Moulin, 1979], and such a profile, if exists, must be pure Nash equilibrium of the original game.

Interestingly, when Moulin introduced this solution concept, he used voting games for that purpose. Indeed, in public information voting games, where explosion of Nash equilibria is a major problem, the elimination of all equilibria but one seems lucrative.

Definition 6.14 (Dominance solvable). A game $\langle f, \boldsymbol{L}\rangle$ is *dominance solvable* if after a finite number of elimination rounds we get that for any player i, all actions are equivalent. A voting rule f is dominance solvable if $\langle f, \boldsymbol{L}\rangle$ is dominance solvable for any preference profile $\boldsymbol{L}$.[7]

[7]Moulin's original definition used cardinal utility profiles.

It is not hard to see that any voting rule where voters have dominant strategies (e.g., dictatorial rules) are dominance solvable (in single elimination round). Moulin focuses on ways to construct more sophisticated dominance solvable rules.

Theorem 6.15 **[Moulin, 1979].** *For any n and m, there is an anonymous and Pareto efficient voting rule that is dominance solvable.*

We do not lay out the full proof, but briefly describe the constructed rule. This is simply a knock-out tournament between candidates, where candidates are matched in some arbitrary pre-defined order, and in each round the less-supported candidate quits.[8]

How about dominance solvability of common voting rules? Dhillon and Lockwood [2004] set out to study this question for the Plurality rule. Clearly, there are profiles $\boldsymbol{L}$ for which $\langle f^{PL}, \boldsymbol{L}\rangle$ is not solvable, and others for which it is. Dhillon and Lockwood gave an almost exact characterization of these profiles. The main observation driving their results is that under Plurality, the *unique* weakly dominated strategy of a voter is to vote for her least preferred candidate. If sufficiently many voters rule out the same candidate, then this candidate cannot win regardless of the other votes, and we can consider a modified game without this candidate. The question is how many voters are "sufficiently many"? For three candidates, there is a crisp answer.

Theorem 6.16 **[Dhillon and Lockwood, 2004].** *Let $Z_c(\boldsymbol{L})$ be the number of voters ranking c last in profile $\boldsymbol{L}$. If $Z_c(\boldsymbol{L}) > \frac{2}{3}n$ for some $c \in A$ then $\langle f^{PL}, \boldsymbol{L}\rangle$ is dominance solvable. If $Z_c(\boldsymbol{L}) < \frac{2}{3}n - 2$ for all $c \in A$ (the bound in the paper is tighter), then $\langle f^{PL}, \boldsymbol{L}\rangle$ is not dominance solvable.*

As we consider games with more candidates, we need greater accordance on the worst candidate (roughly $(1 - \frac{1}{m})n$ voters), then on the worst remaining candidate, and so on. This means that for a reasonable number of candidates, dominance solvable profiles are uncommon. On the other hand, Dhillon and Lockwood show that if a dominance solvable winner exists, it must also be a Condorcet winner.

In Example 6.11, no candidate can be eliminated by dominance. In Example 6.12, types 3 and 5 together compose more than $\frac{2}{3}$ of the voters, and all of them rank a last. Thus $Z_a(\boldsymbol{L}) > \frac{2}{3}n$ and a cannot win and can thus be removed. Once removed, d beats b in a pairwise election, and thus the solution is d.

6.6 EXERCISES

1. Prove that $F^{PAR}(\boldsymbol{L}) \triangleq \{$all Pareto efficient outcomes for $\boldsymbol{L}\}$ is Maskin-monotone and has no veto power. Conclude by Theorem 6.4 that F^{PAR} can be implemented in NE for $n \geq 3$.

2. Let $\boldsymbol{a}$ be a pure Nash equilibrium in Plurality voting with abstentions. Prove that at most one voter votes (all others abstain).

[8] Gretlein [1982, 1983] spotted some gaps in the original proof and explained how to fix them.

3. Consider the following distribution over cardinal types with 3 candidates $A = \{a, b, c\}$, where $x > 0$.

	type 1	type 2	type 3
$U_i(a)$	1	0	-1
$U_i(b)$	0	1	0
$U_i(c)$	-1	-1	x
frequency	0.4	0.4	0.2

Suppose we sample $n = 4$ voters i.i.d. from this distribution. The voting rule is Plurality with uniform tie-breaking. For what values of x do we get that truth-telling is an equilibrium? For what values of x is there a Duverger equilibrium?

4. Suppose there are 5 voters in 2-approval with three candidates $A = \{a, b, c\}$ and uniform random tie-breaking, where 2 votes are $\{a, b\}$ and 3 votes are $\{b, c\}$. Each voter fails to cast her vote with probability ε. What is the probability that each candidate wins? What is the probability of each tie? You are another voter with utilities $(12, 10, 2)$ and weight 0.9 (so you are only pivotal in case of an exact tie). What is the expected utility gain of voting for each of the three strategies?

5. Table 6.3 shows the MMR strategy for type 1 voters with $c < 0.2$ in Example 6.12. Compute the MMR strategy for types 2, 3, and 5. How does it depend on the voting cost c?

CHAPTER 7

Iterative and Sequential Voting

The models in the previous chapter assumed that voters all vote simultaneously, and in particular do not know how others have voted. However, in many realistic scenarios, voters can see the votes of previous voters (recall our Preschool example from the introduction) and may take this information into consideration. We make a distinction between *iterative voting* where voters may revise their vote an unlimited number of times, and *sequential voting* where there is a known procedure with a finite number of steps.

Iterative voting In the iterative voting model, voters have fixed preferences and start from some announcement (e.g., sincerely report their preferences). Votes are aggregated via some predefined rule (e.g., Plurality), but can change their votes after observing the current announcements and outcome. The game proceeds in turns, where a single voter changes his vote at each turn, until no voter has objections and the final outcome is announced. This process is similar to online polls via Doodle or Facebook, where users can log-in at any time and change their vote. Similarly, in offline committees the participants can sometimes ask to change their vote, seeing the current outcome.

The outcome of such a game depends on the exact specification of the voting rule and the process, and also on the strategies that voters use. If we assume voters do not know others' preferences or whether they intend to change their vote, then a plausible assumption is that voters will act in a myopic way. That is, vote in every round as if it is the last one. In game-theoretic terms, each voter will play a best reply to the *current action profile* of the other voters. If no voters wants to change their vote, then by definition the current profile is a PNE. This is the type of behavior we will consider in this chapter. Voters who are more sophisticated on one hand, or have less accurate information about the current state on the other hand, may not follow their best reply and instead use other heuristics. We consider such heuristics in the next chapter.

More on iterative voting appears in Meir [2017], Meir et al. [2017].

For either voting rule and type of behavior we are interested in the following questions.

- Are voters guaranteed to converge, in other words, reach a stable state where no voter wants to move?
- In particular, does such a stable state exist?
- If voters converge, how fast?

- Can we characterize the stable states?
- Is the iterative process leading the society to a socially good outcome?

We show some partial answers for these questions in Sections 7.3 and 7.4.

Sequential voting A general sequential voting mechanism parts the voting process into steps. For example, in each step a single voter may be asked to vote, or all voters may vote on a single issue. After each step, the results become public so voters may take them into account when considering their next steps. The standard solution concept in the game theory literature for such games is *subgame-perfect equilibrium*, which is based on backward induction. We will see in Section 7.5 the application of such solutions to voting games.

7.1 CONVERGENCE AND ACYCLICITY

Local improvement graphs Given a game $\langle f, \boldsymbol{L} \rangle$, a player $i \in N$, and an action profile $\boldsymbol{a} \in \mathcal{A}$, let $I_i(\boldsymbol{a}) \triangleq \{a'_i \in A_i \text{ s.t. } f(\boldsymbol{a}_{-i}, a'_i) \succ_i f(\boldsymbol{a})\}$, that is, the set of all actions that are better replies of agent i in profile $\boldsymbol{a}$.

Any (ordinal or cardinal) game $G = \langle f, \boldsymbol{L} \rangle$ induces a directed graph $H_G = \langle \mathcal{A}, I \rangle$ whose vertices are all action profiles (states) $\mathcal{A}$ and edges I correspond to possible moves by single players. That is, $(\boldsymbol{a}, \boldsymbol{a}') \in I$ if there is a player i such that $a'_i \in I_i(\boldsymbol{a})$, and $\boldsymbol{a}' = (\boldsymbol{a}_{-i}, a'_i)$.

The graph H_G is known as the graph of local improvement steps of G [Andersson et al., 2010, Young, 1993]. In Chapter 8 we will use a similar graph, except the $I_i(\boldsymbol{a})$ will not necessarily be the set of better replies.

The *sinks* of H_G are all states with no outgoing edges. In this chapter, since steps correspond to better replies, a state $\boldsymbol{a} \in \mathcal{A}$ is a sink of H_G if and only if it is a PNE of G. Note that the graph H_G does not uniquely define the outcome of a game. Even if we decide on an initial state, some states may have multiple outgoing edges (due to several actions of the same player or due to different players), and thus there may be many different paths of better replies. See examples in Section 7.2.

Much attention has been given in the game theory literature to the question of *convergence*, and several notions of convergence have been defined based on which paths of the local improvement graph are considered [Apt and Simon, 2012, Kukushkin, 2011, Milchtaich, 1996, Monderer and Shapley, 1996].

Definition 7.1 (Finite improvement property). A game G has the *finite individual improvement property* (we say that G *has FIP*) if the corresponding graph H_G has no cycles.

In other words, any sequence of better replies from any initial state $\boldsymbol{a}^0$ reaches a PNE. Games that have FIP are also known as *acyclic games* and as *generalized ordinal potential games* [Monderer and Shapley, 1996]. Two weaker notions of acyclicity are as follows [Kukushkin, 2011].

- A game G has *weak FIP* if from any initial state $\boldsymbol{a}^0$ there is *some* path in H_G that reaches a PNE. Such games are known as *weakly acyclic.*

- A game G has *restricted FIP* if from any initial state $\boldsymbol{a}^0$ and *any order of players* there is *some* path in H_G that reaches a PNE. We refer to such games as *order-free acyclic.*

Intuitively, restricted FIP means that there is some restriction players can adopt such that convergence is guaranteed regardless of the order in which they play. Equivalently, restricted FIP means that there is some subgraph $H' = \langle \mathcal{A}, I' \rangle$ of the local improvement graph H_G where $I_i(\boldsymbol{a}) \neq \emptyset \Rightarrow I'_i(\boldsymbol{a}) \neq \emptyset$ for all $i \in N, \boldsymbol{a} \in \mathcal{A}$, and H' is acyclic.

Kukushkin identifies a particular restriction of interest, namely restriction to best reply improvements, and defines the *finite best reply property* (FBRP) and its weak and restricted analogs. Note that if player i has some better replies at state $\boldsymbol{a}$, that is, $I_i(\boldsymbol{a}) = S \neq \emptyset$, then some outcome in S must be weakly preferred to all others, and thus $I'_i(\boldsymbol{a}) \neq \emptyset$ even when restricted to best replies.

The *Finite direct reply property* (FDRP) is only relevant for certain voting rules and is defined later on. Figure 7.1 demonstrates entailment relations among the various acyclicity properties.[1]

		FBRP				restricted FBRP	⇒	weak FBRP		
		⇑				⇓		⇓		pure
ordinal potential	⇒	FIP	⇛	FDBRP	⇛	restricted FIP	⇒	weak FIP	⇒	Nash
exists		⇓				⇑		⇑		exists
		FDRP				restricted FDRP	⇒	weak FDRP		

Figure 7.1: A double arrow X ⇒ Y means that any game or game form with the X property also has the Y property. A triple arrow means that any property on the premise side entails all properties on the conclusion side. The bottom row is only relevant for Plurality/Veto, where direct reply is well defined.

We emphasize that the playing agent *must* select an available action, if one exists. For example, we can imagine a dynamic where a voter will only agree to vote for a candidate who is much more preferred than the current winner (say, ranked at least three positions above). Such a voter may not move even though he has available better replies. Thus convergence of this dynamic does not imply restricted FIP.

We say that a game G has *FIP from state* $\boldsymbol{a}$ if all paths from $\boldsymbol{a} \in \mathcal{A}$ reach a PNE. G has *FIP from the truth* if it has FIP from the truthful state $\boldsymbol{a}^* = \boldsymbol{L}$ (for standard rules). We say that a voting rule f has FIP if for *any* preference profile $\boldsymbol{L}$ the induced game $\langle f, \boldsymbol{L} \rangle$ has FIP. The definitions for all other notions of finite improvement properties are analogous.

[1]Compare with the classification of equilibria by their convergence/divergence properties in Section 6.4.2.

7.2 EXAMPLES

This section shows several examples that should help in getting a better understanding of how iterative voting proceeds. Some of these examples are also useful as counterexamples for various properties.

7.2.1 PLURALITY

Example 7.2 **[Meir et al., 2017]** There are three candidates $A = \{a, b, c\}$ and three voters. We have a single fixed voter voting for a whose preferences are irrelevant and who is not playing strategically. The preference profile of the two other voters is defined as $a \succ_1 b \succ_1 c$, $c \succ_2 b \succ_2 a$.

We next analyze the above example, and show that it contains a cycle of best replies. First, we show the game form f^{PL}. Assuming that the action of the third voter remains fixed, we have a 3×3 game where voter 1 is the row player. The scores of all candidates appear in brackets and the winner in curly brackets:

f^{PL}	a	b	c
a	$(3,0,0)\{a\}$	$(2,1,0)\{a\}$	$(2,0,1)\{a\}$
b	$(2,1,0)\{a\}$	$(1,2,0)\{b\}$	$(1,1,1)\{a\}$
c	$(2,0,1)\{a\}$	$(1,1,1)\{a\}$	$(1,0,2)\{c\}$

By also considering the preferences, we can show the full local improvement graph of the game $\langle f^{PL}, \mathbf{L} \rangle$ (scores now omitted). A double arrow indicates an edge of length 2:

$\langle f^{PL}, \mathbf{L} \rangle$	a		b		$\mathbf{c}$
$\mathbf{a}$	$\{a\}*$		$\{a\}*$		$\mathbf{\{a\}}*$
			$\uparrow$		$\uparrow\uparrow$
b	$\{a\}$	$\rightarrow$	$\{b\}$	$\leftarrow$	$\{a\}$
			$\downarrow$		$\uparrow$
c	$\{a\}$	$\rightarrow\rightarrow$	$\{a\}$	$\rightarrow$	$\{c\}$

We can immediately see that the graph $H_{\langle f^{PL}, \mathbf{L} \rangle}$ contains a cycle. Thus the game $\langle f^{PL}, \mathbf{L} \rangle$ in Example 7.2 does not have FIP. This also shows that the Plurality voting rule (the game form f^{PL}) does not have FIP, even though there are three pure Nash equilibria (marked with $*$).

What about the other convergence properties? All arrows in $H_{\langle f^{PL}, \mathbf{L} \rangle}$ correspond to best replies. This means that the game also has a cycle of best replies, and thus $\langle f^{PL}, \mathbf{L} \rangle$ (and Plurality) does not have FBRP.

Note that the truthful profile (marked in bold) is not part of any cycle, so if voters start by telling the truth they are guaranteed to converge (in 0 steps) in this game. In other words, the game $\langle f^{PL}, \boldsymbol{L}\rangle$ has FIP from the truth. Of course, this does not entail that *any* Plurality game has FIP from the truth (see next section).

How about weak acyclicity? We can see that $H_{\langle f^{PL}, \boldsymbol{L}\rangle}$ has only one cycle, and from this cycle there is an edge leading to the sink (a, c). Therefore from each of the nine states there is a path to at least one sink: either there is no path to the cycle, or there is a path to the cycle and from there to (a, c). Therefore the game $\langle f^{PL}, \boldsymbol{L}\rangle$ has weak FIP.

Direct replies We next identify a different restriction, namely *direct reply*, that is well defined under the Plurality rule. Formally, a step $\boldsymbol{a} \xrightarrow{i} \boldsymbol{a}'$ is a direct reply if $f(\boldsymbol{a}') = a'_i$, in other words, if i votes for the new winner.

Note that the edge $(b, b) \to (b, c)$ in Example 7.2 corresponds to a best reply that is not direct, since voter 2 votes for c but the outcome becomes a. Since this is the only cycle in $H_{\langle f^{PL}, \boldsymbol{L}\rangle}$, by removing this edge there are no more cycles, and thus the game $\langle f^{PL}, \boldsymbol{L}\rangle$ has FDRP, and in particular restricted FIP.

7.2.2 VETO

Example 7.3 **[Meir, 2016a]** There are three candidates $A = \{a, b, c\}$ and three voters. One voter is vetoing a and does not actively participate. The two other voters have preferences $b \succ_1 c \succ_1 a$, $a \succ_2 c \succ_2 b$.

The possible actions of each voter are to veto one of the three candidates, hence we get the following game:

$\langle f^{VL}, \boldsymbol{L}\rangle$	$\neg a$	$\boldsymbol{\neg b}$	$\neg c$
$\boldsymbol{\neg a}$	$(0,3,3)\{b\}$	$\boldsymbol{(1,2,3)\{c\}}$ *	$(1,3,2)\{b\}$
$\neg b$	$(1,2,3)\{c\}$	$(2,1,3)\{c\}$	$(2,2,2)\{a\}$
$\neg c$	$(1,3,2)\{b\}$	$(2,2,2)\{a\}$	$(2,3,1)\{b\}$

We can see that the following cycle appears in $H_{\langle f^{VL}, \boldsymbol{L}\rangle}$:

$$(\neg c, \neg b)\{a\} \xrightarrow{1} (\neg b, \neg b)\{c\} \xrightarrow{2} (\neg b, \neg c)\{a\} \xrightarrow{1} (\neg c, \neg c)\{b\} \xrightarrow{1} (\neg c, \neg b)\{a\}.$$

As with Plurality, all edges of the cycle are best replies, which means that $\langle f^{VL}, \boldsymbol{L}\rangle$ (and Veto in general) does not have FBRP or FIP.

Also as with the Plurality example above, $\langle f^{VL}, \boldsymbol{L}\rangle$ does converge if we start from the truthful state $(\neg a, \neg b)$, or restrict attention to direct replies.

Note that a direct reply in Veto is always unique and means to veto the current winner [Lev and Rosenschein, 2012]. Since the first step of the cycle is indirect, and there are no other cycles, the game $\langle f^{VL}, \boldsymbol{L}\rangle$ has FDRP.

7.2.3 BORDA

Unlike Plurality and Veto, for most voting rules the number of actions, and hence the size of the local improvement graph, is exponential in the number of candidates. We next see such an example.

Example 7.4 **[Reyhani and Wilson, 2012]** There are three candidates $A = \{a, b, c\}$ and two voters with preferences $a \succ_1 b \succ_1 c,\ b \succ_2 c \succ_2 a$.

Here is the local improvement graph $H_{\langle f^{BL}, \boldsymbol{L} \rangle}$, where only part of the edges are drawn:

$\langle f^{BL}, \boldsymbol{L} \rangle$	*abc*	*acb*	*bac*	***bca***		*cba*	*cab*
abc	*a*	*a*	*a*	***b***	←	*a*	*a*
				↓		↑	
acb	*a*	*a*	*a*	*a*	→	*c*	*a*
						⇣	
bac	*a*	*a*	*b*	*b*		*b* *	*a*
bca	*a*	*a*	*b*	*b*		*b*	*c*
cba	*a*	*c*	*b*	*b*		*c*	*c*
cab	*a*	*a*	*a*	*c*		*c*	*c*

We can see that there is a cycle, thus Borda does not have FIP, and it can be verified that there are no other cycles in this game. The dashed edge is leaving the cycle into state $(bac, cba)\{b\}$, which has no outgoing edges. Thus $\langle f^{BL}, \boldsymbol{L} \rangle$ has weak FIP and restricted FIP (if we restrict voter 1 to selecting the dashed edge).

However, if we restrict voters' actions to best replies, then all edges of this cycle remain and there are no edges leaving the cycle (note that the dashed edge is not a best reply). Thus $\langle f^{BL}, \boldsymbol{L} \rangle$ (and Borda in general) does not have FBRP or even weak FBRP. [2]

7.3 CONVERGENCE RESULTS

7.3.1 CONVERGENCE IN PLURALITY

We saw in Example 7.2 that restricting voters' actions to direct replies resulted in the elimination of the only cycle. We next show that this is not coincidental.

Theorem 7.5 **[Meir et al., 2017].** *f^{PL} has FDRP. Moreover, any path of direct replies will converge after at most $m^2 n^2$ steps. In particular, Plurality is order-free acyclic.*

[2] In Meir [2016b, 2017] the author overstated the result of Reyhani and Wilson and wrote that their example implies that Borda has no weak FIP (they made no such claim). The more careful analysis above shows that Borda has no weak FBRP, but whether it has weak FIP or restricted FIP remains open. See Table 7.1.

The number of steps until convergence drops to $O(mn)$ if players start from the initial truthful state [Meir et al., 2010] or follow their (unique) direct best reply [Reyhani and Wilson, 2012].

We will demonstrate some of the ideas often used in such proofs, by proving a *weaker* result, namely that any sequence of direct best replies (FDBRP) from the truth converges. Intuitively, when restricted to direct replies from the truth, voters always compromise for less-preferred but more popular candidates, which means that the votes always flow to the more popular candidates. We will see in the next chapter that this idea can be exploited to prove convergence with other behaviors as well.

Proof of FDBRP from the truth. Denote by $w^t = f^{PL}(\boldsymbol{a}^t)$ the winner after step t, and by W^t all candidates that can become winners by at most one additional vote. Denote $a = a_i^{t-1}$. We claim that at any step $a \xrightarrow{i} a_i^t$, the following invariants hold:

(1) $a \neq w^{t-1}$ (the manipulator never leaves the current winner);

(2) a_i^t is i's most preferred candidate in $W^{t-1} \setminus \{a\}$;

(3) $a_i^t = w^t$ (vote goes to the new winner);

(4) $a_i^{t-1} \succ_i a_i^t$ (voter always compromises for a less preferred candidate);

(5) $W^t \subseteq W^{t-1}$ (set of possible winners always shrinks); and

(6) For any voter j, either $a_j^t \notin W^t$, or a_j^t is j's most preferred candidate in W^t.

Assume all of (1)–(6) hold until time $t-1$ and consider step t. We prove by induction that all invariants still hold after step t.

Due to (6), we know that either $a \notin W^{t-1}$, or a is i's most preferred in W^{t-1}. Suppose that $a = w^{t-1}$. Then, we are in the latter case (a is most preferred in W^{t-1}), which means that $w^t = f(\boldsymbol{a}_{-i}^{t-1}, a_i^t) \prec_i a = w^{t-1}$. Thus this cannot be a manipulation step, and $a \neq w^{t-1}$. That is, invariant (1) holds.

Now, since $a \neq w^{t-1}$ the score of the winner after step t does not decrease, only voting candidates in W^{t-1} may change the outcome. Then, invariant (2) follows immediately from our direct best reply assumption.

Invariant (3) follows immediately from the definition of direct replies.

As for (4), either t is the first move of i, in which case $a = \text{top}(L_i) \succ_i a_i^t$, or there had been a step $a' \xrightarrow{i} a$ at some time $t' < t$, in which case a is the most preferred in $W^{t'}$. By inductively applying (5), we have that $W^{t-1} \subseteq W^{t'}$, and thus $a \succ_i c$ for all $c \in W^{t-1} \setminus \{a\}$, and in particular $a \succ_i a'$. Thus (4) holds at step t.

We have $s_{\boldsymbol{a}^t}(w^t) \geq s_{\boldsymbol{a}^t}(w^{t-1}) = s_{\boldsymbol{a}^{t-1}}(w^{t-1})$ (the equality is due to (1)), which means that the score of the winner weakly increased. Thus the threshold to become a possible winner also weakly increased, whereas the score of all $c \neq a_i^t$ weakly decreased. This means that for any $c \neq a_i^t$ we have $c \notin W^{t-1} \Rightarrow c \notin W^t$, and (5) holds at step t.

Invariant (6) holds at $\boldsymbol{a}^0$ by our assumption of truthful initial vote, and continues to hold as long as (5) does, since a more preferred candidate cannot join W^t. Thus (6) holds at step t. Finally, note that by (3), each voter can move at most $m - 1$ times, and thus convergence is achieved in at most $n(m - 1)$ steps. □

This convergence result depends on the tie-breaking. For Plurality games with randomized tie-breaking, order-free convergence is guaranteed from the truth but not from arbitrary states. However, *weak convergence* is guaranteed from any state [Meir et al., 2017].

From the point of view of computational complexity, while verifying whether a given profile $\boldsymbol{a}$ is a PNE is trivial, it is NP-hard to decide whether $\boldsymbol{a}$ is reachable from the truthful profile [Rabinovich et al., 2015]. In contrast, if we assume in addition that voters are truth-biased (see Section 6.3.1), then there is an exact characterization for all reachable equilibria, and they can be verified efficiently. With lazy bias (Section 6.3.2), characterization is even simpler, as any PNE is reachable from the truthful state [Rabinovich et al., 2015].

7.3.2 OTHER VOTING RULES

Veto has similar convergence properties to Plurality, although the proof techniques are different.

Theorem 7.6 [Lev and Rosenschein, 2012, Reyhani and Wilson, 2012]. *Veto has FDRP.*

In contrast, with most common voting rules it is possible to construct examples of cycles, even when voters start by voting truthfully. These results are summarized in Table 7.1.

Further, even variations of the Plurality rule, such as adding voters' weights and/or changing the tie-breaking method, may result in games with cycles. For Plurality with random tie-breaking rule, it can be shown that it is *weakly acyclic*, thereby providing partial explanation to the fact that simulations almost never hit a cycle [Meir et al., 2017]. Whether other common voting rules are also weakly acyclic is an open question, which is particularly of interest for a large number of voters.

One could ask whether there are common (or any) voting rules that have FIP, which even the simple Plurality rule fails to hold. Indeed, beyond the trivial dictatorial rule, one could consider other rules such as the Direct Kingmaker rule (see Section 2.3).

The question of which voting rules, or game forms, have certain convergence properties has attracted attention in the general game-theory literature beyond social choice. In particular, an important open challenge is to characterize all game forms that are FIP. For some contributions in that direction see Kukushkin [2011], Meir et al. [2017].

Table 7.1: All rules in the table use lexicographic tie-breaking. Reference codes: MP+10 [Meir et al., 2010], RW12 [Reyhani and Wilson, 2012], LR12 [Lev and Rosenschein, 2012] (see Lev and Rosenschein [2016] for the full version), L15 [Lev, 2015], M16 [Meir, 2016b], KS+17 [Koolyk et al., 2017], MP+17 [Meir et al., 2017].

Voting rule	FIP	FBRP	FDRP	weak-FBRP	Weak-FIP
Dictator	✓	✓	✓	✓	✓
Plurality	×	× [MP+10]	✓ [MP+17]	✓	✓
Veto	×	× [M16]	✓ [RW12,LR12]	✓	✓
k-approval ($k \geq 2$)	×	× [LR12,L15]	-	×	× [M16]
Borda	×	× [RW12,LR12]	-	× [RW12]	?
PSRs (except k-approval)	×	× [LR12,L15]	-	?	?
Approval	×	× [M16]	-	✓ [M16]	✓
Other common rules	×	× [KS+17]	-	?	?

7.3.3 SIMULTANEOUS MOVES

In practical scenarios, we cannot always expect voters to vote one at a time, especially when there are many. When several voters $S \subseteq N$ change their vote in profile $\boldsymbol{a}$ without coordinating,[3] the resulting profile is $\boldsymbol{a}' = (\boldsymbol{a}_{-S}, \boldsymbol{a}'_S)$ where $\boldsymbol{a}_S = (a'_i)_{i \in S}$ and each $a'_i \in I_i(\boldsymbol{a})$. The sets of group moves we allow induce a different local improvement graph with possibly more edges and cycles.

Unfortunately, even in very simple games it is easy to see that simultaneous moves may add cycles. It is enough to consider Plurality with two voters whose preferences are $c \succ_1 b \succ_1 a$ and $b \succ_2 c \succ_2 a$, and an additional fixed voter for a. Indeed, if both voters move at each turn, they will keep alternating between (b, c) and (c, b).

As it turns out, allowing arbitrary simultaneous moves in *any* game will almost always result in cycles. Clearly if there is no pure Nash equilibrium, then cycles exist even with individual moves.

Theorem 7.7 **[Fabrikant et al., 2010].** *Let G be any game with at least two pure Nash equilibria. Then the graph of group local-improvement steps contains at least one cycle.*

The theorem is very powerful. It entails not just that almost any voting rule f has cycles of simultaneous moves for *some* preferences, but that this is the case for *almost any preference profile* (as we saw that there are typically many pure Nash equilibria)! This may seem like a very negative result, but remember that the existence of cycles only precludes the strongest acyclicity property (similar to FIP, but with simultaneous moves), and weaker acyclicity guarantees may still apply.

Also, the result of Fabrikant et al. applies for games with a finite number of players. Meir [2015] shows that in a *nonatomic* version of Plurality, the game converges even with simultaneous moves by arbitrary subsets of uncoordinated voters.

[3]This is in contrast to *coalitional moves*, which are coordinated. See Gourvès et al. [2016], Kukushkin [2011] and Section 3.4.

7.4 WELFARE IMPLICATIONS

In those cases where an iterative voting game converges, we would like to know how good the outcome is. As with other game-theoretic analyses of voting outcomes, there are at least two different approaches to measure outcome quality:

- with respect to the particular voting rule in question; or
- with respect to an objective measure, such as social welfare, Condorcet efficiency (fraction of profiles where the Condorcet winner was selected out of all profiles with a Condorcet winner) etc.

These two approaches are reflected in previous chapters of the book, for example, when considering Input and Output approximations (Chapter 4) vs. Welfare approximation (Chapter 5).

Dynamic Price of Anarchy A common way to measure the inefficiency in a game due to strategic behavior is the *Price of Anarchy*: the ratio between the quality of the outcome in the worst Nash equilibrium and the optimal outcome [Christodoulou and Koutsoupias, 2005]. In the context of voting, this translates to the question of how far the equilibrium outcome can be from the truthful voting outcome (seeing the truthful outcome as "optimal" according to the voting rule in use).[4]

As we have seen, Nash equilibria in most voting rules can be arbitrarily far from the truth, thus Brânzei et al. [2013] suggested instead to restrict attention to the set of Nash equilibria that are the outcome of some iterative voting procedure starting from the truthful vote.

Consider a score-based voting rule f where the candidate c with the highest score $s(c, \boldsymbol{L})$ wins in preference profile $\boldsymbol{L}$.

Definition 7.8 (Dynamic Price of Anarchy). The *dynamic Price of Anarchy* (DPoA) is defined as

$$DPoA(f) = \min_{\boldsymbol{L} \in \mathcal{L}(A)^n} \min_{\boldsymbol{L}^* \in EQ^T(f, \boldsymbol{L})} \frac{s(f(\boldsymbol{L}^*), \boldsymbol{L})}{s(f(\boldsymbol{L}), \boldsymbol{L})} ,$$

where $EQ^T(f, \boldsymbol{L})$ is the set of all profiles $\boldsymbol{L}^*$ s.t.:

- there is a path of best replies from the truthful profile $\boldsymbol{L}$ to $\boldsymbol{L}^*$; and
- $\boldsymbol{L}^*$ is a Nash equilibrium of $\langle f, \boldsymbol{L} \rangle$.

Brânzei et al. [2013] show that the DPoA of Plurality is at most 1 (see Exercise 7.6(4) for a somewhat weaker claim); and that the DPoA in Veto depends on the number of candidates m. In particular for any fixed m the DPoA in Veto is constant, regardless of n. The DPoA in Borda, on the other hand, is $\Omega(n)$, meaning that equilibria can be arbitrarily bad even for a fixed number of alternatives.

[4]This notion of approximation is similar to Output approximation, but instead of comparing the outcome under two different rules, we consider the same rule under truthful and rational behavior.

Objective quality metrics The fact that we chose to use a particular voting rule does not necessarily mean that this rule represents the optimal outcome for every profile. The selection of the rule might be affected by other properties of the rule such as simplicity, tradition, and so on. We may thus have multiple criteria for a "good outcome" and ask how well a given voting rule satisfies them. Note that we focus on the equilibria outcomes rather than on the truthful outcome.

For example, we may be interested in the social welfare of the voters (as measured by some linear or other utility scale), in the likelihood of finding the Condorcet winner when one exists, or avoiding the Condorcet loser, and so on.

This question was studied using extensive simulations in Meir et al. [2014] for the Plurality rule, where it was shown that equilibria outcomes are *better* than the truthful outcome under most metrics observed. Koolyk et al. [2017] performed similar simulations for several other voting rules, and obtained mixed results w.r.t. the social welfare. More interestingly, rules that are not Condorcet consistent (Bucklin, STV) are more likely to find the Condorcet winner under rational play than under truthful voting, whereas the Condorcet efficiency of Condorcet-consistent rules only slightly declines.

That said, when the number of voters is large, the initial outcome is almost always an equilibrium (whether truthful or not), and thus the effect of best-reply dynamics becomes negligible. We consider the question of welfare again after introducing heuristic voting in Section 8.3.

7.5 SEQUENTIAL VOTING

Sequential voting is similar to iterative voting in the sense that there are multiple rounds instead of a single concurrent decision. However, unlike iterative voting, the entire structure of the game is known in advance. Thus sophisticated players can plan ahead and predict the future actions of other rational players.

7.5.1 SUBGAME PERFECT EQUILIBRIUM

One of the earliest formal analyses of strategic voting is due to Farquharson [1969]. Notably, this brilliant text precedes even the G-S theorem and has inspired much of the work on social choice to this day. We will only bring here one of the models suggested in Farquharson's essay.

Recall that the problems with strategic voting stem from the need to select among three or more strategies. When only two actions are available, voters have a dominant strategy in almost any reasonable voting rule. It makes sense, therefore, to break the decision voters need to make into a sequence of binary choices. For example, such a decision could be on which of two specified candidates should be eliminated (as in the knockout tournament from Theorem 6.15). More generally, a *binary voting procedure* is a binary tree whose leaves are candidates (each candidate appears at least once), and in each round there is a majority vote (starting from the root) on which branch to take.

Given such a tree, it becomes clear what voters should do at the last round of voting when both children are leaves. Once this is clear, there is only one reasonable outcome to every decision in the penultimate round, and thus voters' actions at this round can be predicted as well. We can continue and solve the entire game by backward induction. This behavior was termed as *sophisticated voting* (the solution of this process in known in the game theory literature as *subgame perfect equilibrium*. See, e.g., Leyton-Brown and Shoham [2008, p. 35]). In contrast, in *sincere voting*, each voter votes according to, her preferences at every round in favor of the branch containing the most favorable candidate (voting rules may vary in how a sincere voter would break ties when the favorite candidate appears in both branches).

Consider Example 6.11 and suppose that in the first round voters should vote on $\{a, b\}$ vs. $\{c, d\}$. If voters are strategic, they will first realize that in the second round b must beat a, and d must beat c. Thus in the first round, types 2, 4, and 5 will vote for $\{c, d\}$. In contrast, if voters are truthful, only types 4 and 5 will vote for $\{c, d\}$ (although in this case d will win regardless).

More recent papers analyzed the outcome that results from a specific breakdown to such binary decisions. McKelvey and Niemi [1978] are perhaps most responsible for the popularity of Farquharson's model. With some modifications, they proved the following.[5]

Theorem 7.9 **[McKelvey and Niemi, 1978].** *If $a \in A$ is a Condorcet winner in $\mathbf{L}$, then under any binary voting procedure and sophisticated voting, a will be the winner.*

In contrast, they show that a Condorcet winner may not be elected when voters are truthful (Exercise 6.6(6)).

A natural setting to use sequential voting is when voters have combinatorial preferences over a set of k issues (so that the candidates are all 2^k combinations, assuming issues are binary). In that case, we get a binary voting procedure where in each level of the tree a single issue is voted on, and each candidate appears in exactly one leaf. Xia et al. [2011] studied restrictions on voters' preferences such that the backward-induction outcome coincides with the truthful outcome.

Sequential participation Desmedt and Elkind [2010] considered a sequential voting game where in each round a single voter votes, rather than all voters voting on a single binary decision. Clearly, if there are only two candidates and no voting cost, then all voters will be truthful and the truthful outcome will be elected. However, as we saw earlier, if voting carries even a negligible cost, then some voters may choose not to participate.

Desmedt and Elkind characterize the behavior equilibrium of strategic voters with negligible positive participation cost and show that in any subgame perfect equilibrium, the truthful outcome will be elected, even if some voters will choose to abstain. Further, the margin between the two candidates will be at most 1.

[5]Farquharson assumed that voters can report their entire strategy in advance, which leads to multiplicity of equilibria. McKelvey and Niemi considered a variation where voting in every node is done sequentially, which is equivalent to subgame perfection. This results in unique equilibrium strategies that can be derived by the standard backward induction procedure.

With more than two candidates (plus the possibility of abstention), it is difficult to make overreaching statements on the structure of equilibrium, and the paper provides examples where some voters vote strategically in equilibrium and the outcome depends on the order of voters. Still, Desmedt and Elkind design algorithms that can compute a subgame perfect equilibrium in time that is polynomial in *one of* m and n, but exponential in the other.

7.5.2 ITERATED MAJORITY VOTING

Recall that when there are only two alternatives, any monotone voting rule is clearly strategyproof and thus voting is straightforward. Airiau and Endriss [2009] suggest a sequential game where starting from some initial alternative ("incumbent"), we run a finite sequence of majority votes (or some other rule based on quota), where in each round a voter that is selected at random may suggest a competing alternative to the incumbent (an "usurper"). If the suggestion is accepted, the usurper becomes the new incumbent, and either way the game proceeds to the next iteration. In contrast to iterative voting where voters react myopically, here voters only care about the outcome of the final iteration. As lengthy definitions are required for accurate presentation, we demonstrate the spirit of the model with an example.

Suppose there are 4 alternatives $A = \{a, b, c, d\}$ and 3 voters, with utility vectors $U_1 = (4, 2, 1, 0)$, $U_2 = (1, 4, 3, 0)$, and $U_3 = (2, 1, 4, 0)$. We think of $\boldsymbol{U} = \boldsymbol{U}^0$ as a $m \times n$ *utility matrix* whose i'th column is U_i. Note that if voters vote truthfully, then in a pairwise match, a will beat b, b will beat c, c will beat a, and all candidates will beat d.

Suppose that the current winner is a, and there is one last iteration to play (thus truthful voting is a dominant strategy). The only way that the winner will change, is if c is the suggested usurper. This will happen if either voter 2 or 3 is selected. Note that even though voter 2 prefers b over c, he will not suggest b as an usurper, as b would fail to beat a. Thus the probability of transition in the last step from a to c is $\frac{2}{3}$, and with probability $\frac{1}{3}$ (when voter 1 is selected), the outcome will remain a. We can similarly compute the transition probabilities in case the incumbent is b, c or d, which results in the stochastic transition matrix W^1 (row x contains transition probabilities from incumbent x, and sums to 1).

$$\boldsymbol{U}^0 = \begin{pmatrix} 4 & 1 & 2 \\ 2 & 4 & 1 \\ 1 & 3 & 4 \\ 0 & 0 & 0 \end{pmatrix} \qquad W^1 = \begin{pmatrix} \frac{1}{3} & 0 & \frac{2}{3} & 0 \\ \frac{2}{3} & \frac{1}{3} & 0 & 0 \\ 0 & \frac{2}{3} & \frac{1}{3} & 0 \\ \frac{1}{3} & \frac{1}{3} & \frac{1}{3} & 0 \end{pmatrix} \qquad \boldsymbol{U}^1 = W^1 \cdot \boldsymbol{U}^0 = \frac{1}{3} \begin{pmatrix} 6 & 7 & 10 \\ 10 & 6 & 5 \\ 5 & 11 & 6 \\ 7 & 8 & 7 \end{pmatrix}$$

Next, suppose that there are two iterations left. How would the voters vote? Column i of the matrix $\boldsymbol{U}^1 = W^1 \cdot \boldsymbol{U}^0$ shows the expected utility of each candidate to voter i as an incumbent of the next iteration. Note for example that the utility of d to voter 1 is $\frac{7}{3}$, since if d is the incumbent of the penultimate iteration, the final outcome is equally likely to be any of a, b and c. From $\boldsymbol{U}^1$, we can derive all pairwise majorities in the current iteration (which happen to be transitive): $d \succ a \succ c \succ b$. We can then conclude that if a is the incumbent, d will be proposed and selected w.p. of $\frac{2}{3}$, and otherwise a remains. We can go on and compute all transition probabilities in the current iteration and the expected utilities to be used in the next step.

$$W^2 = \begin{pmatrix} \frac{1}{3} & 0 & 0 & \frac{2}{3} \\ 0 & \frac{1}{3} & 0 & \frac{2}{3} \\ 0 & 0 & \frac{1}{3} & \frac{2}{3} \\ 0 & 0 & 0 & 1 \end{pmatrix} \qquad \boldsymbol{U}^2 = W^2 \cdot \boldsymbol{U}^1 = \frac{1}{9}\begin{pmatrix} 20 & 23 & 24 \\ 24 & 22 & 19 \\ 19 & 27 & 20 \\ 21 & 24 & 21 \end{pmatrix}$$

Another step will show that $W^3 = W^2$. When there are t steps remaining, compute $\boldsymbol{U}^{t-1}$ and W^t by backward induction as above. Interestingly, there is a positive probability for candidates like d, which have no chance of winning and are undesired by all voters, to be incumbents during part of the game.

The main question studied in Airiau and Endriss [2009] is *convergence*. They define several notions of convergence for the sequence of utility matrices $\boldsymbol{U}^0, \boldsymbol{U}^1, \ldots, \boldsymbol{U}^t, \ldots$. *Intrastate convergence* means that the sequence has a limit $\boldsymbol{U}^*$; *interstate convergence* means that all rows of U^* are identical; and *fundamental convergence* means that in the limit all rows of W^t are identical. In other words, fundamental convergence means that for a sufficiently long sequence of steps there is a stationary distribution over winners, and interstate convergence means that all candidates have the same utility for i as usurpers—this is the expected utility for voter i from the next incumbent in a sufficiently long game (regardless of the initial state). The process is similar to a Markov chain, except that the transition matrix in every iteration is different.

Airiau and Endriss show that fundamental convergence implies interstate convergence, which in turn implies intrastate convergence. Then they study theoretically and empirically some sufficient conditions for convergence. In particular, they show empirically (and analytically for $m = 2$) that setting the quota to strictly under 50% always results in interstate convergence. In contrast, with a quota above 60%, convergence rapidly becomes rare.

The paper also shows an example where a Condorcet winner exists but is not guaranteed to win even if the number of iterations is large.

7.6 EXERCISES

1. Consider a profile with three candidates $\{a, b, c\}$ and three voters whose preferences are $L_1 = a \succ b \succ c$, $L_2 = b \succ c \succ a$, and $L_3 = a \succ c \succ b$. Draw $H_{\langle f^{PL}, \boldsymbol{L}\rangle}$ and $H_{\langle f^{VL}, \boldsymbol{L}\rangle}$. Which convergence properties do they have, if any?

2. Give examples of local improvement graphs with the following properties: (a) FIP; (b) restricted FIP but not FIP; (c) weak FIP but not restricted FIP; (d) has a sink but not weak FIP. Can you define games whose induced graphs have these properties?

3. Show an example of a preference profile in 2-Approval that contains a cycle of better replies.

4. In this question we eliminate one unnecessary condition in the proof of Theorem 7.5 and use it to prove that $DPoA(f^{PL}) \leq 1$. Consider Plurality with lexicographic tie-breaking and assume that voters *start from the truthful state*. (a) Prove that best replies are also direct; (b) conclude that the invariants in the proof of Theorem 7.5 hold without the restriction to direct replies; and (c) prove that a candidate whose initial Plurality score is strictly lower than $sw^0 - 1$ cannot win under best reply dynamics.

5. Suppose that 5 voters are voting over 3 binary issues, in other words, $A = \{0, 1\}^3$. A sequential Majority rule is used to determine the outcome on all issues. Assume that preferences are public information, and that voters follow the backward induction solution. Give an example (i.e., a preference profile) where changing the order of issues changes the elected outcome. Is it possible that a Condorcet loser gets elected?

6. Construct an example of a binary voting tree where a Condorcet winner exists but is not selected if voters are truthful.

CHAPTER 8

Voting Heuristics

The equilibrium analysis and the best-reply dynamics considered in the previous chapters make the following implicit assumptions.

- Voters know exactly how other voters currently vote, or a distribution thereof.
- Voters are myopic: always vote as if the game ends after the current turn.
- Voters always vote in a way that improves or maximizes their utility.

In this last technical chapter we will relax some or all of these assumptions and consider models that aim to capture more realistic voting behavior. We then ask what will happen if all voters behave in a given way. In particular, Section 8.2 mirrors the analysis of Section 7.3 regarding convergence to equilibrium of various heuristics. Section 8.3 considers the societal implications when voters apply some of the heuristic strategies in the chapter, mainly via computer simulations.

8.1 HEURISTIC VOTING MODELS

8.1.1 AD HOC HEURISTICS

Most heuristics are similar to best reply in that they only assume the voter knows her own preferences and has some information about the current voting profile (e.g., the score of each candidate, or their current ranking). However, in contrast to best or better reply, a heuristic step may or may not change the outcome. It thus reflects the belief of the voter that she might be pivotal even if this is not apparent from the current state. In the next subsection we will look more closely at such a rational (or bounded-rational) justification, but for now we will be satisfied with just describing some heuristics that have been proposed.

In the following description, we assume f is some score-based rule, unless specified otherwise. Let L_i be the real preferences of voter i, and $\boldsymbol{a} = (a_1, \ldots, a_n)$ be the *current* action profile and $\boldsymbol{s} = (s_1, \ldots, s_m)$ be the current scores of all candidates derived from $\boldsymbol{a}$. We denote by c_j the candidate with the j'th highest score in $\boldsymbol{s}$.

Crucially, some of these heuristics depend on some internal private parameters, which can be used to explain behavioral differences among voters.

- "k-pragmatist" [Reijngoud and Endriss, 2012]: Here, each voter has a parameter k_i. The voter considers her most preferred candidate among $W = \{c_1, \ldots, c_{k_i}\}$ and puts it at the top of L'_i (ranking all other candidates below according to L_i).

- "Threshold": Similar to k-pragmatist, except instead of a fixed parameter k, the set of "possible winners" W consists of all candidates whose score s_j is above some threshold $T_i(\boldsymbol{a})$ (i.e., the threshold may depend both on i and on the current state).
- "M1" or "Second chance" [Grandi et al., 2013]: If the current winner is not i's best or second-best choice according to L_i, she moves her second-best alternative to the top position.

Consider Example 6.11 with the Borda voting rule and $n = 100$. Thus candidates' scores are $(55, 175, 175, 195)$. A type 1 voter that is 2-pragmatist will move b (which is the more preferred candidate among $W = \{b, d\}$) to the top, and vote $L'_1 = bacd$. A type 1 voter with a threshold of 110 will act the same. A type 3 voter will not change her vote if she is a 2-pragmatist, but if she follows Second chance, she will move c to the top and vote $L'_3 = cbda$.

The other heuristics below are various restrictions of better reply or best reply. As such, they can only expand the set of equilibria in a given voting game.

- "M2" or "Best upgrade" [Grandi et al., 2013]: This is a restriction of best reply to candidates that are ranked above the current winner $f(\boldsymbol{a})$ in the current vote a_i.
- "Upgrade" [Obraztsova et al., 2015]: Similar to Best Upgrade, except the upgraded candidate is not necessarily placed first (only high enough to win).
- "Unit upgrade" [Obraztsova et al., 2015]: Similar to Upgrade, except the upgraded candidate is moved exactly one step up (if this is enough to win).
- "Top" [Koolyk et al., 2017]: A restriction of best reply, to actions that place the new winner at the top of L'_i.
- "TopBottom" [Koolyk et al., 2017]: Similar to Top, and in addition places the current winner at the bottom of L'_i.
- "KT" [Koolyk et al., 2017]: A restriction of best reply that minimizes the Kendall-Tau distance (or "swap distance") between L_i and L'_i.

Note that all of the above heuristics (and most of the other heuristics we consider in the chapter) are *myopic* in the following sense: Each heuristic can be written as a function whose input is the voting rule f, the current state $\boldsymbol{a}$, and the voter's preferences L_i, and whose output is a new vote $a'_i \in A_i$ or a set of allowed votes $A'_i \subseteq A_i$.

Heuristics based on previous actions One may also consider more sophisticated heuristics that are based on recent history or the entire history. One such class of heuristics was suggested by Bowman et al. [2014], who considered iterative voting over multiple binary issues. They assume that voters compute three independent factors for each alternative: its *utility* (which is fixed); its *attainability* (which depends on how much support it has in the current action profile);

and its *learning factor*, which depends on the empirical utility of voting for this alternative in previous rounds.

Approval voting and Laslier's Leader rule Some voting rules enable the use of specific heuristics. Laslier [2009] suggested the following heuristic strategy for Approval voting, called the *Leader rule*:

Algorithm 8.2 The Leader rule

Input: preferences L_i, scores $\boldsymbol{s}$.
Initialize a'_i to the empty vote
$c_1 = \text{argmax}_{j \in A} s_j$ //the Leader
$c_2 = \text{argmax}_{j \in A \setminus \{c_1\}} s_j$ //the Runnerup
Approve $\{c \in A : c \succ_i c_1\}$ in a'_i
if $c_1 \succ_i c_2$
  Approve c_1 in a'_i
Return a'_i

Consider Example 6.11 with $n = 100$ where in the initial state each voter votes only for her top candidate. The leader is d with 35 votes and the runner-up is b with 30. A type 3 voter would approve (after observing the initial state) candidates b and c, which he prefers to the leader d. A type 4 voter would approve candidate c, which is preferred to the leader d, and also the leader d, which is preferred to the runner-up b. Type 5 voters will only approve d, type 1 will approve a, b, c and type 2 will approve a, d. In the resulting profile, the score vector is $(10, 35, 60, 65)$.

The Leader rule has some attractive properties. First, it requires very little information on the other votes, it does not require any computation or non-trivial inference from the voter, and it is *sincere*. That is, a voter never approves a candidate that she likes less than some other disapproved candidate.[1] In addition, the Leader rule can be justified as a rational behavior. For example, if we assume that there are small perturbations in the ballot, as in Section 6.5.2, then voting according to the Leader rule maximizes the voter's expected utility. This is true for any cardinal utility U_i that fits L_i, and continues to hold under different noise models [Laslier, 2009]. The crucial requirement under which this holds is that a tie with the leader is substantially more likely than any other tie, and thus it is strictly beneficial to approve candidates preferred to the leader.

A different heuristic that was suggested exclusively for Approval is *Average Target* [Merrill, 1981], which assumes the voter will approve all candidates whose value is above the expected utility (which of course requires some statistical information or belief, as in Section 6.4.1).

[1] Note that as in Approval, the voters do not report a preference order: the concept of *truthful voting* does not apply to it.

Laslier's Borda heuristics (LBH) Laslier [2010b] also suggests a specialized heuristic for Borda (although it can be applied to any voting rule): First consider the leader c_1 and the lowest ranked candidate c_m according to the poll $\boldsymbol{s}$, and place the more preferred one at the top of the reported vote L'_i, and the other at the bottom. Then consider each candidate in the order they are ranked in $\boldsymbol{s}$, and place each $c_2, \ldots, c_{m-1}$ either at the bottom or at the top of the available slots in L'_i according to whether $c_1 \succ_i c_j$ or $c_j \succ_i c_1$, respectively.

8.1.2 LOCAL DOMINANCE

A different approach to derive heuristic voting behavior, is to consider a formal way to model voter's *uncertainty* regarding the outcome. Then, based on her beliefs, the voter selects the action (ballot) that is best for her.

One way to generate such models is to take one of the "rational" models, for example, from Section 6.4, and relax the assumption that the beliefs on which the voters set their optimal action are accurate, assuming instead beliefs are derived in some heuristic way. For example, Merrill [1981] suggests exactly this on the tie probabilities p_{xy} in Equation (6.1) (Myerson and Weber [1993] later derive these probabilities from the actual strategic actions of all voters).

In this section though, we consider models of voters with *bounded rationality*. In contrast with the models we considered in Section 6.4, voters may not assign exact probabilities to outcomes (real or estimated), and in particular cannot compute expected utilities.[2]

Local dominance [Meir et al., 2014] explicitly separates the beliefs of the voters on candidates' scores and their strategic actions. The belief is represented by a set of possible states S, which contains all score vectors "close" to the current profile $\boldsymbol{a}$. The strategic action is based on dominance relations with the set of possible states. The dominance relation is given in Definition 8.1, then the construction of S, and the application for a heuristic voting step in Definition 8.2. The definitions apply for any voting rule based on scores, including non-standard rules such as Approval, where the set of actions A_i is not necessarily $\mathcal{L}(A)$.

Definition 8.1 (Local dominance). Let f be a scoring-based rule, and L_i preferences of voter i. Given two actions $a'_i, a_i \in A_i$ and a subset of states S, we say that a'_i *locally dominates* a_i *in* S if:

- $f(\boldsymbol{s}', a'_i) \succeq_i f(\boldsymbol{s}', a_i)$ for all states $\boldsymbol{s}' \in S$; and
- $f(\boldsymbol{s}'', a'_i) \succ_i f(\boldsymbol{s}'', a_i)$ for at least one state $\boldsymbol{s}'' \in S$.

It is not hard to see that for any S, local dominance is a transitive and anti-symmetric relation. Note that if S contains all possible score vectors of $n-1$ voters, then local dominance coincides with global dominance (Definition 2.13). If S contains a single score vector $\boldsymbol{s}'$ induced

[2]Also note that expected utility is undefined for a voter with ordinal preferences, even if we had such a distribution.

by $\boldsymbol{a}_{-i}$, then local dominance coincides with better reply (Definition 2.10). In general, a state s' does not have to be induced by an action profile of $n-1$ voters.

In this section, we assume that the set $S = S_i(\boldsymbol{a})$ is derived from $\boldsymbol{a}_{-i}$ using the ℓ_∞ distance metric and the voter's uncertainty parameter r_i. An alternative metric is considered in Example 8.4(2).

$$S_i(\boldsymbol{a}) \triangleq \{\boldsymbol{s}' \in [n]_0^m : \forall c \in A, |s_c^{-i} - s'_c| \le r_i\}.$$

The following heuristic applies for any scoring-based voting rule and any way that S is derived. Note that a local dominance step may not be unique.

Definition 8.2 (Local dominance step). A *local dominance step* for a voter i at state $\boldsymbol{a} = (\boldsymbol{a}_{-i}, a_i)$ is a vote $a'_i \in A_i$ that locally dominates a_i in $S_i(\boldsymbol{a})$ and is not locally dominated by another vote.

This behavior encodes bounded rationality under *loss aversion*: the voter will make a strategic move only if certain (according to her beliefs) that this move will not hurt her, and might be beneficial for her.

Local dominance in Plurality Plurality is easier to analyze than other rules, since the set of possible actions A_i coincides with the set of candidates A. Recall our running example (Example 6.11) with $n = 100$. A type 1 voter under Plurality with $r_i = 8$ considers as possible all outcomes $(s_a, s_b, s_c, s_d) \in S_i$, where $s_a \in [9-8, 9+8] = [1, 17]$, $s_b \in [22, 38]$, $s_c \in [17, 33]$, and $s_d \in [27, 43]$. Note that there is no possible state where a wins. However, there is a state $\boldsymbol{s}' \in S_i$ (for example $\boldsymbol{s}' = (9, 23, 30, 31)$) where $f(\boldsymbol{s}', a) = d$ but $f(\boldsymbol{s}', c) = c \succ_i d$. Thus, voting $a'_i = c$ locally dominates $a_i = a$ in S_i. A similar calculation shows that $a'_i = b$ also locally dominates a, so voter i has two possible moves from the truthful profile: $A'_i = \{b, c\}$.

Let $w = f^{PL}(\boldsymbol{a})$, and denote by $W_i \triangleq \{c \in A : s_c^{-i} > s_w^{-i} - 2r_i - 1\}$ all candidates that can beat w with $2r_i + 1$ additional votes.

Observation 8.3 $c \in W_i$ if and only if c is a possible winner, in other words, if there is a state $\boldsymbol{s}' \in S_i(\boldsymbol{a})$ such that $f^{PL}(\boldsymbol{s}', c) = c$.

To see why, consider the state $\boldsymbol{s}' \in S_i(\boldsymbol{a})$ where $s'_c = s_c^{-i} + r_i$ and $s'_a = s_a^{-i} - r_i$ for all $a \neq c$. Note that even for two voters with $r_i = r_j$, it is possible that $W_i \neq W_j$ since $S_i \neq S_j$.

Worst-Case Regret minimization (WCR) [Meir, 2015] and Non-Myopic voting (NM) [Obraztsova et al., 2016a] are similar in the way they derive the set of possible winners W_i but then make some different behavioral assumptions on action selection. Regret minimization in voting was described in Section 6.5.1, where it was shown to lead to truthful voting. The main difference in the WCR model of Meir [2015] is that the voter only considers the set S_i of possible world states.

It turns out that local dominance provides a (bounded) rational justification to the threshold heuristics we described above, at least in Plurality.

Lemma 8.4 [Meir et al., 2014] *Suppose that $a_i \notin W_i$. $a_i \xrightarrow{i} a'_i$ is a local dominance step in Plurality if and only if: $|W_i| \geq 2$, and a'_i is the most-preferred candidate in W_i.*

Proof sketch. Clearly if $|W_i| = 1$, voter i cannot affect the outcome, and thus no vote dominates any other vote.

When $|W_i| \geq 2$, denote by a^* and b^* the most preferred and least preferred candidates in W_i, respectively. Consider a state $\mathbf{s}' \in S_i$ where b^* wins but any other candidate in W_i can beat b^* with a single additional vote. In state $\mathbf{s}'$, voting for a^* is strictly better for i than voting for any other candidate, and voting for any candidate in $W_i \setminus \{b^*\}$ is strictly better than voting for b^*.

To establish dominance, we also need to show that there is no state where voting a_i is better than voting for a'_i. Indeed, since $a_i \notin W_i$, for any state $\hat{\mathbf{s}} \in S_i$ voting for $a'_i \in W_i$ either does not change the outcome ($f(\hat{\mathbf{s}}, a'_i) = f(\mathbf{a})$), or changes the outcome to a'_i. If $a'_i = a^*$ this never hurts i, thus $a'_i = a^*$ locally dominates a_i. In contrast, consider the state $s'' \in S_i$ where a^* wins but any other candidate in W_i can beat a^* with a single additional vote. In state $\mathbf{s}''$ voting for any $a'_i \in W_i \setminus \{a^*\}$ changes the outcome to $f(s'', a'_i) = a'_i \prec_i a^* = f(\mathbf{a})$. Thus a'_i does not locally dominate a_i.

Finally, suppose that some $a'_i \neq a^*$ also locally dominates a_i. If $a'_i \in W_i$, then consider a state where a^* wins unless i votes for a'_i (thus a'_i does not dominate a_i). If $a'_i \notin W_i$, it is dominated by a^*, thus $a_i \xrightarrow{i} a'_i$ is not a valid local dominance move.

For a complete proof, one has to address the various cases of tie-breaking. □

If $a_i \in W_i$, then a_i it may still be dominated, but only if a_i is the least preferred in the set (see Example 8.4(1)).

Therefore, a voter that simply follows the threshold heuristics by voting for her most preferred candidate in W_i is essentially strategizing according to local dominance.[3] If the voter selects a WCR step rather than following local dominance moves, then this coincides with the threshold heuristics exactly [Meir, 2015]. See also Exercise 7.6(ch7,exe5).

8.1.3 OTHER DOMINANCE-BASED HEURISTICS

A distance metric is just one way to define a set of "possible states," and it makes sense mainly for rules based on scores. Other ways have also been proposed where the set of states S_i is derived from the voting profile $\mathbf{a}$ by keeping only some of the information. For example, voters may

[3] Note that while any LD move is consistent with the threshold heuristics, the converse does not always hold. For example, if there are five possible winners above the threshold, then a move from the third-preferred to the most preferred is not a local dominance move. Indeed, a loss-averse voter would refrain from such a dangerous move.

only know who is the prospective winner, meaning that S_i contains all profiles $\boldsymbol{a}'$ where $f(\boldsymbol{a}') = f(\boldsymbol{a}_{-i})$ (note that this is possible with any voting rule f). Other examples are knowledge only of the candidates' true ranking (which only applies for scoring rules), or the pairwise majority relations among them [Endriss et al., 2016, Reijngoud and Endriss, 2012]. The belief structure can also be an arbitrary partition of the state space [Conitzer et al., 2011, van Ditmarsch et al., 2013], which is the common way to model knowledge in modal logic and in partial information games.

In all of these papers (as in Section 8.1.2), it is assumed that a voter will manipulate using strategy a_i' only if her current vote a_i is weakly dominated by a_i' in the set $S_i(\boldsymbol{a})$. We note that for some belief structures this may lead to triviality. For example, f^{PL} is Plurality and $S_i(\boldsymbol{a})$ contains all profiles A^{n-1} (i.e., voters have no information at all), then the only weakly dominated strategy is to vote for the worst candidate in L_i. In particular the truthful vote is never dominated, and Plurality becomes "strategyproof." Thus to obtain interesting results, one needs to be careful in selecting the belief structure. Endriss et al. [2016] shows the following implications of the information voters have: A voter may have a dominance manipulation in Plurality (and in many other voting rules) even if only the identity of the winner is known; If the voter knows all pairwise majority relations, then there are dominance manipulations in Borda and Copland, but not in k-Approval; Maximin can also be manipulated given this information, but only if there is a (weak) Condorcet winner. Compare this approach with the approaches presented in Chapter 4 to circumvent the G-S theorem.

Optimism A voter's uncertainty might not be objective but influenced by her own preferences. For example, the voter may anticipate a higher possible score to favorite candidates. Two such models based on voter's optimism were suggested in Obraztsova et al. [2016a], Reyhani et al. [2012]. In both models, voters with greater uncertainty will consider a larger set of candidates as possible winners and will vote strategically to one of them.

Multiple uncertainty levels In a recent paper, Lev et al. [2018] extend the local dominance framework by adding multiple levels of uncertainty. In the context of voting, this means that a voter first considers the states that are most likely to occur and rules out dominated actions. If there are still several available actions, then the next set of (less likely states) are considered and so on. They show that some of the ad hoc heuristics in Section 8.1.1 can be justified as "dominance moves" in such a model with an appropriate belief structure.

8.2 EQUILIBRIUM AND CONVERGENCE

A common way to describe Nash equilibrium is as a "state where all agents prefer to keep their current action, assuming everybody else does." However, the formal definition we use (Definition 2.12) does not make any explicit assumptions on agents' expectations. It simply requires that every agent plays a best reply to the current action profile. The rationality assumption is cap-

Table 8.1: Positive results means FIP from the truth for uniform population (where relevant). Convergence for other conditions was not studied. (* common PSRs including Borda and Plurality). Reference codes: RE12 [Reijngoud and Endriss, 2012], GL+13 [Grandi et al., 2013], OM+15 [Obraztsova et al., 2015].

Voting rule	k-pragmatist	Second Chance	Best upgrade	Upgrade	Unit Upgrade
PSRs	✓ [RE12]	✓	✓ [GL+13]	?	✓ [OM+15] *
Maximin	✓ [RE12]	✓	✓ [GL+13]	✓ [OM+15]	✓ [OM+15]
Copeland	✓ [RE12]	✓	✓ [GL+13]	?	?
Bucklin	-	✓	?	?	✓ [OM+15]
all rules	-	✓ [GL+13]	?	?	?

tured by the requirement to "best reply," and we can naturally extend the equilibrium concept to agents with bounded rationality by replacing "best reply" with any other heuristics.

Given a voting game $\langle f, \boldsymbol{L} \rangle$ and a population of voters with well-defined heuristics, a *voting equilibrium* is simply a profile of valid votes $\boldsymbol{a}$, such that for each $i \in N$, the heuristic action of i in profile $\boldsymbol{a}$ is to keep her current action a_i. Then a Nash equilibrium is a special case of voting equilibrium when voters' heuristics are to play a better (or best) reply at any state $\boldsymbol{a}$.

Given a voting rule and a heuristic, two important questions are: (a) does an equilibrium exist? and (b) will voters converge to an equilibrium? The latter question can be further divided according to the various convergence properties discussed in Section 7.1.

Recall that FIP means convergence is guaranteed regardless of the initial state, the order of voters, and which available reply they choose. As any heuristic simply replaces the (possibly empty) set of better replies with some other set, we can modify the definition of FIP, or FIP from the truth, accordingly.

Convergence was studied in several recent papers. Some heuristics are very easy to analyze. For example, when starting from the truthful vote, then voters using the k-pragmatist or the Second Chance heuristics will move at most once [Grandi et al., 2013, Reijngoud and Endriss, 2012]. Therefore, FIP from the truth is immediate.

Obraztsova et al. [2015] identified some common structure for heuristic dynamics, which can be used to prove convergence for various combinations of voting rules and heuristics. All these studies are restricted to voters that start by reporting the truth and use exactly the same heuristics. Results are summarized in Table 8.1.

The Leader rule Laslier [2009] characterizes the equilibria with a large population of voters following the Leader rule in Approval. Informally, he shows that the score of the winner c^* will be the number of voters preferring c^* to the runner-up, and that the score of any other candidate c will be the number of voters preferring c to c^*. As a consequence, if a Condorcet winner exists, it will always be selected in equilibrium.

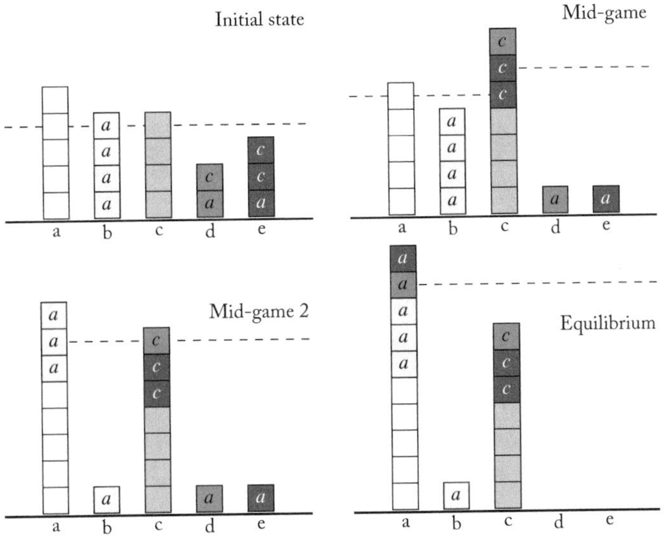

Figure 8.1: Convergence under Local Dominance. The top left figure shows the initial (truthful) state of the game. The letter inside a voter is his, second preference. The dashed line marks the threshold of possible winners W_i for voters of type $r_i = 2$. For example, since a beats b in tie-breaking, b needs 2 more votes to win in the initial state. Note that due to tie-breaking, the threshold is not the same for all candidates. In the next two figures we can see voters leaving their candidates (who are not possible winners for them) to join one of the leaders. The last figure shows an equilibrium that was reached. Figure based on Meir et al. [2014].

Uncertainty-based heuristics Uncertainty-based heuristics are more involved, especially when the society is composed of voters with different uncertainty levels. Therefore, they have been studied mostly for the Plurality rule. On the other hand, it turns out that the Local Dominance heuristics have very strong convergence properties. An example of a Plurality game where voters use the Local Dominance heuristics is in Figure 8.1.

Theorem 8.5 [Meir, 2015, Meir et al., 2014]. *Plurality with the Local Dominance heuristics is FIP. This holds for any population of voters with diverse uncertainty levels* $r_i > 0$.

When considering voters with the same uncertainty parameter $r_i = r \geq 0$ starting from the truth, the proof is very similar to that of Theorem 7.5, and some of the proofs in Table 8.1: votes always transfer from less popular candidates to more popular ones, and voters always compromise for less-preferred but more popular candidates. The reason we get FIP and not FDRP is that whenever there is slight uncertainty ($r_i > 0$), the requirement that a'_i is itself undominated rules out "indirect" replies to candidates outside W_i.

Limited convergence properties were also shown for other uncertainty-based heuristics. We summarize them in Table 8.1.

Table 8.2: Convergence results for Local Dominance and Worst-Case Regret minimization. All results are for Plurality. MLR14 [Meir et al., 2014], M15 [Meir, 2015], OL+16 [Obraztsova et al., 2016a].

Heuristics	Population	FIP	FIP from truth	equilibrium exists
LD	uniform	✓ [M15]	✓ [MLR14]	✓
LD	uniform + truth bias	?	✓ [MLR14]	✓
LD	diverse	✓ [M15]	✓ [M15]	✓
WCR	uniform	?	✓ [M15]	✓
WCR	diverse	×	×	× [M15]
NM	uniform	?	✓ [OL+16]	✓
NM	diverse	×	× [OL+16]	?

A weaker type of convergence is to a *stable winner* rather than a stable state, that is, to a set of profiles where voters may still move between, but in all of them the winner is the same.

Theorem 8.6 [Endriss et al., 2016]. *Copland and Maximin with dominance moves that use only information about the winner, converge to a stable winner from the truthful state.*

This result is an "FIP style" convergence result, with the specified caveats. Endriss et al. also provide an "FDRP style" result for convergence of PSRs, where the restriction they impose is selecting a dominance manipulation that minimizes the Kendal-tau distance to the current ballot. Interestingly, such a restriction *does not* imply convergence of best reply for PSRs [Koolyk et al., 2017]. That is, the limited information about the poll scores is essential for a positive convergence result.

8.2.1 SAMPLING EQUILIBRIUM

Suppose that each voter only observes a small finite sample from the population and votes strategically based on this sample [Osborne and Rubinstein, 2003]. The paper does not assume that players necessarily optimize their response by Bayesian or other reasoning, and allows for a wide range of strategies and heuristics. They demonstrate the usefulness of this solution concept by analyzing a three-candidate game where voters have single-peaked preferences and observe a sample of either two or three votes. Voters may apply an arbitrary strategy that depends on their preferences and observations. Osborne and Rubinstein characterize the unique equilibrium that emerges from a given heuristic strategy.

A nice thing about the Sampling equilibrium model is that it allows heterogeneous voting behavior even among voters of the same type (i.e., same preferences and same heuristics) and explains why even unpopular candidates may retain some support. We demonstrate the model using the following example: there are three candidates $\{a, b, c\}$ under the Plurality voting rule. Ninety percent of the voters are non-strategic, of which 40% vote a, 36.8% vote b, and 13.2% vote c. The remaining 10% prefer $c \succ b \succ a$ and apply the following heuristic strategy: they ask

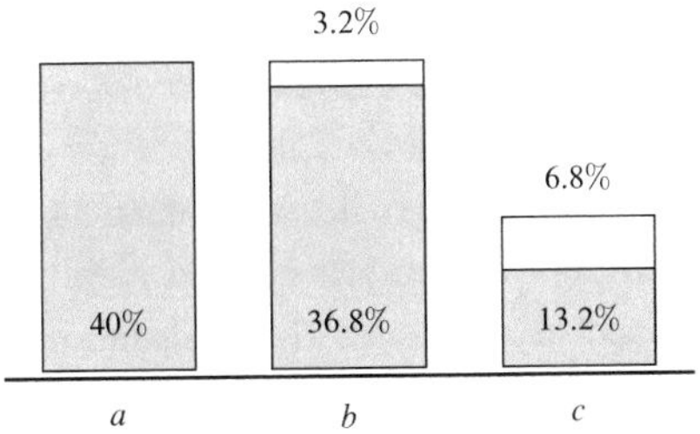

Figure 8.2: An example of sampling equilibrium in Plurality. The gray voters are fixed. The white voters all have preferences $c \succ b \succ a$.

two random voters for their actions. If the sample is $\{a, b\}$ (indicating that a tie between in a and b is likely) then they compromise and vote for b. In any other case they remain truthful and vote c (which can be justified, for example, by truth bias).

How will the sampling equilibrium look like? Suppose first that all strategic voters are truthful. Since there is a non-zero probability of getting a sample $\{a, b\}$, some of them must change their action. On the other hand, it cannot be that all of them vote for b because the probability of a sample different than $\{a, b\}$ is also non-zero. In equilibrium, the voters will split so that 3.2% vote b and the remaining 6.8% vote c (see Figure 8.2). The probability of the sample $\{a, b\}$ is exactly

$$2\frac{s_a}{s_a + s_b + s_c}\frac{s_b}{s_a + s_b + s_c} = 2 \cdot 0.4 \cdot (0.368 + 0.032) = 2 \cdot 0.4 \cdot 0.4 = 0.32.$$

That is, a fraction of 0.32 from the 10% of strategic voters will choose to compromise, so this is indeed an equilibrium. Moreover, voters will converge to this equilibrium from any initial state (in the terminology of Section 6.4.2; this is a GESE).

The model also offers a natural way to combine heuristics that make sense for a small number of voters (like best reply and its refinements in Section 8.1.1) and apply them to a large population. It also makes a clever use of probabilities without assigning cardinal utilities.

In Lu et al. [2012], there is an in-depth analysis of the computational and other challenges for a voter who wants to vote in the optimal way given samples of the other votes. However, without further assumptions and restrictions of the model, it is hard to say anything about equilibrium.

8.3 IMPLICATIONS OF HEURISTIC VOTING

For most notions of heuristic behaviors and equilibrium concepts, it is possible to tailor various edge cases that demonstrate extreme and possibly unrepresentative outcomes. Computer simulations are an important tool that lets us study the common implications and the average effect of various parameters in the model. Simulations are carried out by generating preference profiles from a *culture*, in other words, some distribution of preferences (e.g., Impartial culture, Urn [Berg, 1985], Plackett-Luce [Guiver and Snelson, 2009], etc.), setting the initial profile (truthful or other), and then letting voters make a heuristic move according to the model in question. For some heuristics, voters' parameters should also be decided or sampled up front. The total amount of free parameters available for the designer of the simulation is huge, making it difficult to compare different works. Still, one can try to derive conclusions from patterns that repeat across many different settings.

Heuristic reaction to a poll We start with simulations that assume very simple heuristics that are based on a public poll. That is, alternatives are ranked based on the true preference profile and a given voting rule, each voter reacts to the poll once, and we compute the winner again. Laslier [2010b] compared the outcome under three voting rules with the following heuristics: Plurality (2-pragmatist), Borda (LBH); and Approval (Leader rule). Laslier starts from the truthful profile (for Approval, from the truthful Copeland profile), and records the outcome after each of the first five strategic iterations, where in each iteration *all* voters react. One finding is that Borda is very unstable with the defined behavior, with every iteration radically changing the outcome. See Figure 8.3. As for the outcome, Plurality typically performs the worst. Borda is beneficial to the society when voters are truthful, but both social welfare and Condorcet efficiency deteriorate due to strategic behavior in most cultures. Approval emerges from these simulations as the rule that is most stable and most likely to select high quality outcomes.

Lehtinen [2010] considered a model where voters vote once after getting a noisy signal on the true preferences of the other voters. Voters each calculate pivot probabilities somewhat similarly to the calculus of voting models but are bounded rational in two ways: they only reason a single iteration forward and do not reach an equilibrium, and their confidence levels may not reflect the true strength of the signal. Lehtinen performed extensive simulations with three alternatives, varying the strength of the signal, the correlation between voters' preferences, heterogeneity in the population, and other parameters. One of the conclusions was that in most of these scenarios, strategic behavior increases social welfare in Plurality[4] but decreases social welfare in Approval (see Figure 8.4). Yet the welfare in Approval is almost always better than in Plurality, with or without strategic behavior. Lehtinen discusses the results at length and provides several possible explanations. Note that the results on Approval stand in contrast to those of Laslier [2010b]. Since results are robust across multiple cultures in both settings, the difference is likely due to the different information available to the voters and the different heuristics the authors consider.

Bowman et al. [2014] ran simulations of iterative voting on three-issue agendas (i.e., eight alternatives), using the heuristics described in Section 8.1.1. They concluded that the results after many iterations outperform those of simultaneous voting.

[4]Lehtinen also shows that when only some types engage in strategic behavior, this mainly benefits the *other types*, who are not strategic.

$(n = 11)$	$K = 3$	$K = 5$	$K = 15$
Pr. Condorcet :	1	1	1
Rule		(Condorcet)	
Plurality	.736	.528	.305
Copeland	1	1	1
AV1	1	1	1
AV2	1	1	1
AV3	1	1	1
AV4	1	1	1
AV5	1	1	1
Borda	.868	.777	.616
Borda 1	.850	.508	.158
Borda 2	.996	.557	.166
Borda 3	.983	.732	.223
Borda 4	.999	.739	.271
Borda 5	.998	.806	.340

$(n = 11)$	$K = 3$	$K = 5$	$K = 15$
Pr. Condorcet :	.989	.966	.833
Rule		(Copeland)	
Plurality	.782	.538	.233
Copeland	1	1	1
AV1	.989	.966	.833
AV2	.989	.966	.833
AV3	.989	.998	.935
AV4	.989	.967	.860
AV5	.989	.966	.841
Borda	.874	.538	.624
Borda 1	.861	..479	.122
Borda 2	.988	.500	.133
Borda 3	.966	.661	.316
Borda 4	.990	.674	.393
Borda 5	.965	.753	.476

Figure 8.3: The effect of heuristic strategies on Condorcet efficiency in two different cultures. K is the number of alternatives. Figures based on Laslier [2010b].

Equilibrium It should first be noted that in practice, convergence to equilibrium is achieved (in simulations) almost always, whether or not this is guaranteed by theorems. Moreover, this convergence is typically very quick. This also holds when voters gain their information on the other votes through a social network [Tsang and Larson, 2016].

The ad hoc heuristics we mentioned typically lead to a better winner in terms of Condorcet efficiency and social welfare [Grandi et al., 2013]. As with the best reply simulations mentioned in Section 7.4, heuristics that are restrictions of best reply, such as M2, usually result in a mild improvement: they only trigger a small amount of strategic behavior, and thus in many profiles the equilibrium is simply the initial state. We can see however that the effect of some heuristics and in particular 2-pragmatist on Plurality voting is substantial; see Figure 8.5.

Extensive simulations of Plurality voting with Local Dominance heuristics show the following [Meir et al., 2014].

- As the uniform uncertainty level r increases, there is more strategic interaction among voters (more moves) until a certain point, from which strategic interaction declines.
- With diverse uncertainty levels, there is more strategic interaction than with uniform uncertainty.

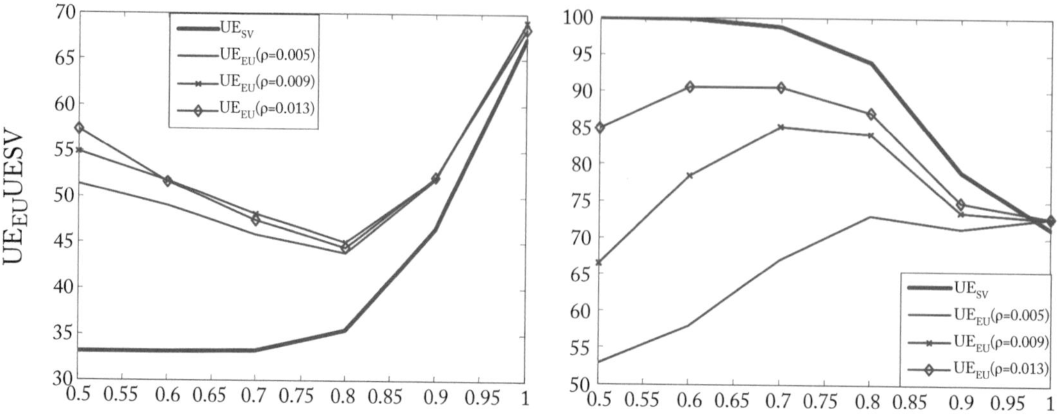

Figure 8.4: The effect of heuristic strategies on voters' social welfare in Plurality (left) and Approval (right). The C parameter (X-axis) indicates the correlation between voters' types and the intensity of their cardinal utility, where for $C = 1$ cardinal values are sampled independently of the ordinal preferences. The Y-axis measures the social welfare. The thick line is the truthful baseline. The thin lines show the welfare when half of the voters (sampled at random) are strategic for three different signal strengths. Figures based on Lehtinen [2010].

- With more strategic interaction, the welfare measures (including social welfare, Condorcet consistency, and others) tend to improve, reaching a significant improvement around the peak of strategic activity; see Figure 8.6 (top)
- With more strategic interaction, votes are more concentrated around two candidates (Duverger Law); see Figure 8.6 (bottom).

These findings are consistent among a broad class of preference distributions and for different numbers of voters and candidates. Therefore at least for Plurality, it seems that equilibria reached under Local Dominance resemble outcomes we observe in reality, and avoids unreasonable or highly inefficient Nash equilibria.

8.4 EXERCISES

1. Consider local dominance moves under Plurality, and suppose that $|W_i| \geq 2, r_i \geq 1$. Prove that if the current vote a_i is in W_i, then it is locally dominated if and only if it is the least preferred candidate in W_i. What if $r_i = 0$?

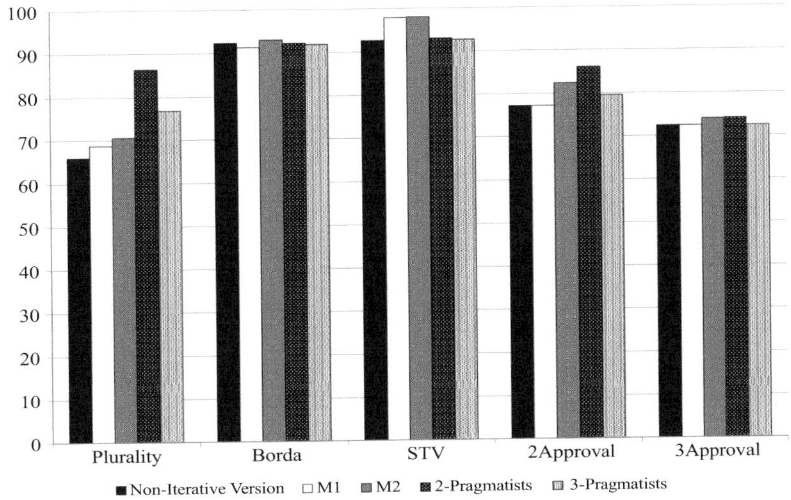

Figure 8.5: Condorcet efficiency under various voting rules and ad hoc heuristics. Simulations use a population of $n = 50$ voters voting over $m = 5$ alternatives. Preferences were sampled from the impartial culture distribution. Figure based on Grandi et al. [2013].

2. In this question we consider local dominance moves but change the distance metric used to define the set of possible states $S_i(\boldsymbol{a})$. Suppose we derive S by using the ℓ_1 metric instead of ℓ_∞:

$$\hat{S}_i(\boldsymbol{a}) = \{\boldsymbol{s}' \in [n]_0^m : \sum_{c \in A} |s_c^{-i} - s'_c| \leq 2r_i\}.$$

(i) Show an example where $a_i \in W_i$, is not the least preferred and is still locally dominated, that is, one direction of Exercise (1) does not hold. (ii) Show that the other direction still holds. (iii) Prove that local dominance moves still converge from the truth when all voters have the same r_i.

3. Consider the three-approval rule, where voters start from the truthful state and follow the 3-pragmatist heuristics. Prove that the game must converge.

4. Consider the following preference profile under Approval.

type	1	2	3
preferences	*abcde*	*bcdea*	*cdeab*
number of voters	40	50	30

- Suppose that in the initial profile, each voter approves her top two candidates. In each iteration, all voters revote according to the Leader rule. What is the outcome after a single iteration? Assume lexicographic tie breaking.

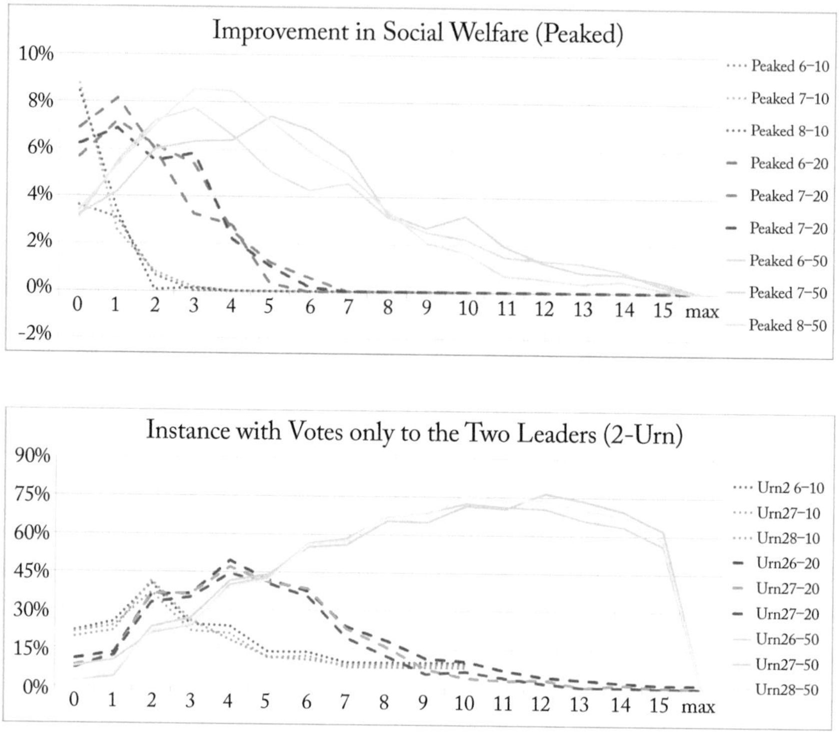

Figure 8.6: The effect of local dominance voting on equilibrium outcome. The X-axis is the uncertainty parameter r (same for all voters). Preference profiles are sampled from various distributions: *peaked* means a distribution over all single-peaked profiles where the pick is selected uniformly. *Urn-2* is a Polya-Eggenberger urn distribution with two urns. The first and second number next to each entry in the legend specify the number of alternatives and number of voters, respectively. Top: Relative increase in social welfare (under linear cardinal utilities). Bottom: fraction of profiles in which all voters ended up voting for only two candidates. Figures based on Meir et al. [2014].

- Show that this dynamic leads to a cycle.
- Is there a pure equilibrium for this game? Prove your answer.

5. In the *worst case regret* (WCR) dynamics under Plurality $f = f^{PL}$, a voter with cardinal utilities considers the set of possible states S as in Local Dominance, and votes for the candidate minimizing the regret

$$a_i^* = \operatorname*{argmin}_{a \in A} \max_{s \in S} \max_{b \in A} (U_i(f(s,b)) - U_i(f(s,a))).$$

   (a) Show that a_i^* is the most preferred candidate in W_i (see Observation 8.3).

(b) Show an example (voters may have difference uncertainty paramters) where the WCR dynamics does not converge.

CHAPTER 9

Summary: Toward a Complete Theory of Strategic Voting

9.1 EMPIRICAL AND EXPERIMENTAL FINDINGS

The material covered in this book was limited to formal definitions and mathematical theorems, with the exception of some computer simulations. However, to better understand strategic voting, theoretical models must be confronted with, and influenced by, empirical data and behavioral experiments.

In this short section we briefly mention some insights on the way actual people vote strategically, based on empirical and experimental work. The purpose is not to provide a full coverage of this literature (which would require another book) but to list several phenomena that often occur in the real world, and thus plausible theories of voting should take into account.

One such phenomenon that has been discussed in the book already is Duverger's Law, which asserts that under Plurality voting, almost all votes will go to only two candidates. There is some empirical evidence that this indeed occurs in actual political voting and plenty of discussion on how to correctly interpret the empirical results [Benoit, 2002, Reed, 1990].

Evidence for strategic voting behavior Political scientists have been inferring strategic voting patterns in various countries [Alvarez et al., 2006, van Deemen and Vergunst, 1998] and institutions [Chamberlin et al., 1984, Hall, 1992] and study how prevalent strategic voting is in practice.

For example, Blais and Nadeau [1996] developed a methodology for measuring strategic voting and applied it to the 1988 Canadian election. They show that the propensity to vote strategically increases with the difference in utility that voters assign to their first and second candidates. Strategic voting also increases when the most-preferred choice is perceived as having lower chances of winning and when there is a close race between less-preferred candidates. The latter two findings have also been shown in primary election studies [Abramson et al., 1992] and in lab experiments [Tal et al., 2015, Tyszler and Schram, 2016, Van der Straeten et al., 2010]. Evidence from lab experiments suggest that voters tend to manipulate more in Plurality than in other rules [Bassi, 2015]. Other researchers ran experiments with more sophisticated multi-round votes and got mixed results [Herzberg and Wilson, 1988].

Fisher [2004] compares several measurement techniques of strategic (or "tactical") voting using empirical election data.

Rational voting and Heuristics When testing how various rational and heuristic models fit the actual individual votes in lab experiments, Van der Straeten et al. [2010] found that rational models (such as the calculus of voting in Section 6.4) and ad hoc heuristics like k-pragmatist (see Section 8.1.1) are at par with about 66% accuracy, whereas truthful voting only has 54% accuracy. In 2-round Plurality, the 3-pragmatist heuristics has 70% accuracy, which is much higher than other tested strategies. In STV, *truthful* voting accurately predicts 90% of the votes.

In Approval, the Leader rule (see Section 8.1.1) accurately predicts 88% of the votes of human voters. As Laslier [2010a] points out, simpler heuristics such as "approve the top two candidates" do almost as well but cannot explain results in real-world voting where the number of approvals varies between voters and between campaigns.

Part of the difficulty in reaching better accuracy with a single strategic model is explained by the substantial interpersonal differences between voters [Tal et al., 2015].

Voter heterogeneity was also pointed out by Blais et al. [2000], who compared the "calculus of voting" model (which is based on expected utility maximization) to empirical election data. They found that the factors in the model do explain some of the behavior, but for many voters the "sense of duty" plays a more important role.

Equilibrium and convergence Bassi [2015], Forsythe et al. [1996] ran controlled experiments where groups of voters play repeatedly with the same preferences in the Plurality and Borda rules. Players were more likely to converge in Plurality, often to a profile that is close to the equilibrium of the MW model. Experiments with iterative voting are less common but suggest that voters converge after few rounds [Tal et al., 2015].

Both the findings of Tal et al. [2015] (where voters voted in a round-robin order) and those of Van der Straeten et al. [2010] (where in each round all voters revoted) suggest that the individual voter uses a myopic strategy that takes into account the current votes of others rather than trying to predict or influence future votes.

Kearns et al. [2009] ran controlled experiments where voters had an exogenous incentive to reach consensus (but also individual preferences on the two available alternatives), and their information about the votes of others was mediated by a social network structure. They show that the topological features of the network have strong effect on convergence, where in some topologies the minority opinion can prevail.

Welfare implications Forsythe et al. [1996] concluded that the Borda voting rule was preferable to Plurality, as the Condorcet loser alternative wins the election less frequently in Borda. Bassi Bassi [2015] favored the plurality voting system, which, although considered the most manipulable, yielded higher social welfare.

The experiments in Van der Straeten et al. [2010] used profiles that were crafted to imitate various political situations. In some, they show that only under Approval do the voters reach a socially desirable outcome. Regenwetter and Grofman [1998] also favor Approval voting and showed in an empirical study that the Condorcet winner would always be selected, if exists.

Further, the outcome in all datasets coincides with the Borda winner, despite the fact that voters did not report their full preferences.[1]

Voter turnout We mentioned in Section 6.3.2 the *paradox of voting*, which under some models predicts low voter turnout. Voter turnout in political voting is affected by a plethora of personal, social, and legal considerations [Jackman, 1987, Powell, 1986]. These are not easily captured in a game-theoretic framework, and thus less useful for evaluating game-theoretic models.

Levine and Palfrey [2007] measured factors that affect voter turnout in lab experiments with two candidates, so there is no strategic decision to make other than vote/abstain. As in Bassi [2015], Forsythe et al. [1996], subjects played the same game repeatedly for many rounds. In particular, they showed that turnout is higher in smaller groups, when the race is close, and among the supporters of the minority candidate. They also showed that the individual behavior is not deterministic, and consistent with quantal response equilibrium.

Bandwagon effect This term, also known as "herding", relates to a situation where voters add their votes to the leading candidate. There is empirical evidence for strong bandwagon effect in political voting [Bartels, 1988, Nadeau et al., 1993], Wikipedia promotions [Leskovec et al., 2010], and other domains. Herding has also been shown in *controlled experiments* where the voters get no information on candidates' quality from their position in the polls: Tal et al. [2015] showed that about half of the human voters in a lab experiment consistently vote for their second-ranked candidate when that candidate is leading the poll, and that the propensity to vote in this way increases with the gap from the next candidates.

Social utilities In political voting, both voters' preferences and behavior is often similar to that of their peers [Riecken, 1959]. An empirical study on votes in the meeting scheduling platform `doodle.com` [Reinecke et al., 2013] shows that voters are more likely to approve slots that are popular, possibly to take into account the preferences of others. More surprisingly, they are also likely to approve *highly unpopular* slots, possibly to appear more cooperative to others [Zou et al., 2015].

9.2 CONCLUSION

Social choice is perhaps the oldest topic that has been analyzed as a strategic game, much before the term game theory was coined. Yet while economists, political scientists, mathematicians, and recently computer scientists, all agree that Nash equilibrium is not an appropriate solution concept for voting, there does not seem to be a single acceptable theory.

This might be due to the fact that, as in other cases that concern human behavior, strategic voting involves many factors. Some of these factors may be domain-specific and/or depend on complex cognitive and social processes. The desiderata that appear in Section 6.1 may pro-

[1] Methods used to construct voters' hypothesized linear preferences (see Falmagne and Regenwetter [1996]) could be useful in other studies.

vide a way to compare the strengths and weaknesses of the many different theories: Some make implausible informational or cognitive assumptions, some predict behavior that is contrary to empirical evidence, some are prohibitively difficult to analyze, and some are very convincing when restricted to specific scenarios. Similarly, under standard game-theoretic assumptions, many mechanism design problems hit impossibility results that force us to use approximation. Comparing mechanism performance based on different notions of approximation (such as input approximation, output approximation and welfare approximation) may help to flash out situations in which we would prefer each mechanism.

I sincerely hope that this book will both encourage and enable researchers to develop new and better models that will improve our understanding of strategic voting. This, in turn, may lead to the design of improved aggregation mechanisms that lead the society to outcomes that are better for everyone.

— Reshef Meir, Haifa 2018

Bibliography

P. R. Abramson, J. H. Aldrich, P. Paolino, and D. W. Rohde. Sophisticated voting in the 1988 presidential primaries. *American Political Science Review*, 86(1):55–69, 1992. DOI: 10.2307/1964015. 127

D. Abreu and A. Sen. Subgame perfect implementation: A necessary and almost sufficient condition. *Journal of Economic Theory*, 50(2):285–299, 1990. DOI: 10.1016/0022-0531(90)90003-3. 76

D. Abreu and A. Sen. Virtual implementation in Nash equilibrium. *Econometrica: Journal of the Econometric Society*, 59(4):997–1021, 1991. DOI: 10.2307/2938171. 76

S. Airiau and U. Endriss. Iterated majority voting. In *Proc. of the 1st International Conference on Algorithmic Decision Theory (ADT)*, pages 38–49, 2009. DOI: 10.1007/978-3-642-04428-1_4. 105, 106

N. Alon, M. Feldman, A. D. Procaccia, and M. Tennenholtz. Strategyproof approximation of the minimax on networks. *Mathematics of Operations Research*, 35(3):513–526, 2010. DOI: 10.1287/moor.1100.0457. 61

R. M. Alvarez, F. J. Boehmke, and J. Nagler. Strategic voting in British elections. *Electoral Studies*, 25(1):1–19, 2006. DOI: 10.1016/j.electstud.2005.02.008. 127

D. Andersson, V. Gurvich, and T. D. Hansen. On acyclicity of games with cycles. *Discrete Applied Mathematics*, 158(10):1049–1063, 2010. DOI: 10.1016/j.dam.2010.02.006. 94

E. Anshelevich and J. Postl. Randomized social choice functions under metric preferences. *Journal of Artificial Intelligence Research*, 58:797–827, 2017. DOI: 10.1613/jair.5340. 57

K. R. Apt and S. Simon. A classification of weakly acyclic games. In *Proc. of the 5th Symposium on Algorithmic Game Theory (SAGT)*, pages 1–12, 2012. DOI: 10.1007/978-3-642-33996-7_1. 94

H. Aziz, F. Brandt, and M. Brill. On the tradeoff between economic efficiency and strategyproofness in randomized social choice. In *Proc. of the 12th International Conference on Autonomous Agents and Multi-Agent Systems (AAMAS)*, pages 455–462, 2013. 40

S. Bade and Y. A. Gonczarowski, Gibbard-Satterthwaite success stories and obvious strategyproofness. In *Proc. of the 2017 ACM Conference on Economics and Computation*, pages 565–565, 2017. 31

S. Barberà. Strategyproof social choice. In K. arrow, A. Sen, and K. Suzumura, Eds., *Handbook of Social Choice and Welfare*, vol. 2, pages 731–831, 2011. 18

S. Barberà and B. Dutta. General, direct and self-implementation of social choice functions via protective equilibria. *Mathematical Social Sciences*, 11(2):109–127, 1986. DOI: 10.1016/0165-4896(86)90020-x. 76

S. Barberà, J. Massó, and A. Neme. Voting under constraints. *journal of Economic Theory*, 76(2):298–321, 1997. DOI: 10.1006/jeth.1997.2301. 31

S. Barberà, A. Bogomolnaia, and H. Van Der Stel. Strategyproof probabilistic rules for expected utility maximizers. *Mathematical Social Sciences*, 35(2):89–103, 1998. DOI: 10.1016/s0165-4896(97)00043-7. 40

L. M. Bartels. *Presidential Primaries and the Dynamics of Public Choice*, Princeton University Press, 1988. 129

J. Bartholdi and J. Orlin. Single Transferable Vote resists strategic voting. *Social Choice and Welfare*, 8:341–354, 1991. DOI: 10.1007/bf00183045. 36

J. Bartholdi, C. A. Tovey, and M. A. Trick. The computational difficulty of manipulating an election. *Social Choice and Welfare*, 6:227–241, 1989. DOI: 10.1007/bf00295861. 35, 36

A. Bassi. Voting systems and strategic manipulation: An experimental study. *Journal of Theoretical Politics*, 27(1):58–85, 2015. DOI: 10.1177/0951629813514300. 127, 128, 129

K. Benoit. The endogeneity problem in electoral studies: A critical re-examination of Duverger's mechanical effect. *Electoral Studies*, 21(1):35–46, 2002. DOI: 10.1016/s0261-3794(00)00033-0. 127

S. Berg. Paradox of voting under an urn model: The effect of homogeneity. *Public Choice*, 47(2):377–387, 1985. DOI: 10.1007/bf00127533. 22, 120

N. Betzler, R. Niedermeier, and G. J. Woeginger. Unweighted coalitional manipulation under the borda rule is NP-hard. In *Proc. of the 22nd International Joint Conference on Artificial Intelligence (IJCAI)*, pages 55–60, 2011. 38

E. Birrell and R. Pass. Approximately strategyproof voting. In *Proc. of the 22nd International Joint Conference on Artificial Intelligence (IJCAI)*, pages 67–72, 2011. DOI: 10.21236/ada582553. 44

D. Black. On the rationale of group decision-making. *Journal of Political Economy*, 56(1):23–34, 1948. DOI: 10.1086/256633. 31

A. Blais and R. Nadeau. Measuring strategic voting: A two-step procedure. *Electoral Studies*, 15(1):39–52, 1996. DOI: 10.1016/0261-3794(94)00014-x. 127

A. Blais, R. Young, and M. Lapp. The calculus of voting: An empirical test. *European Journal of Political Research*, 37(2):181–201, 2000. DOI: 10.1111/1475-6765.00509. 128

A. Bogomolnaia and H. Moulin. A new solution to the random assignment problem. *Journal of Economic Theory*, 100(2):295–328, 2001. DOI: 10.1006/jeth.2000.2710. 40

A. Bogomolnaia, H. Moulin, and R. Stong. Collective choice under dichotomous preferences. *Journal of Economic Theory*, 122(2):165–184, 2005. DOI: 10.1016/j.jet.2004.05.005. 40

T. Börgers. Undominated strategies and coordination in normalform games. *Social Choice and Welfare*, 8(1):65–78, 1991. DOI: 10.1007/bf00182448. 75

C. Boutilier, I. Caragiannis, S. Haber, T. Lu, A. D. Procaccia, and O. Sheffet. Optimal social choice functions: A utilitarian view. *Artificial Intelligence*, 227:190–213, 2015. DOI: 10.1016/j.artint.2015.06.003. 54

C. Bowman, J. K. Hodge, and A. Yu. The potential of iterative voting to solve the separability problem in referendum elections. *Theory and Decision*, 77(1):111–124, 2014. DOI: 10.1007/s11238-013-9383-2. 110, 120

S. J. Brams and P. C. Fishburn. Approval voting. *American Political Science Review*, 72(3):831–847, 1978. DOI: 10.2307/1955105. 33, 34

F. Brandl, F. Brandt, and J. Hofbauer. Incentives for participation and abstention in probabilistic social choice. In *Proc. of the 14th International Conference on Autonomous Agents and Multi-Agent Systems (AAMAS)*, pages 1411–1419, 2015. 80

F. Brandl, F. Brandt, M. Eberl, and C. Geist. Proving the incompatibility of efficiency and strategyproofness via SMT solving. *Journal of the ACM*, 65(2):6, 2018. DOI: 10.1145/3125642. 40

F. Brandt, V. Conitzer, U. Endriss, J. Lang, and A. D. Procaccia, Eds. *Handbook of Computational Social Choice*, Cambridge University Press, 2016. DOI: 10.1017/cbo9781107446984. xvi, 38, 47

S. Brânzei, I. Caragiannis, J. Morgenstern, and A. D. Procaccia. How bad is selfish voting? In *Proc. of the 27th Conference on Artificial Intelligence (AAAI)*, 2013. 102

G. Carroll. A complexity result for undominated-strategy implementation. *Technical Report*, Mimeo, 2014. 76

R. Cavallo. Optimal decision-making with minimal waste: Strategyproof redistribution of VCG payments. In *Proc. of the 5th International Conference on Autonomous Agents and Multi-Agent Systems (AAMAS)*, pages 882–889, 2006. DOI: 10.1145/1160633.1160790. 51

J. R. Chamberlin, J. L. Cohen, and C. H. Coombs. Social choice observed: Five presidential elections of the American psychological association. *The Journal of Politics*, 46(02):479–502, 1984. DOI: 10.2307/2130971. 127

Y. Cheng and S. Zhou. A survey on approximation mechanism design without money for facility games. In *Advances in Global Optimization*, pages 117–128. 2015. DOI: 10.1007/978-3-319-08377-3_13. 61

Y. Cheng, W. Yu, and G. Zhang. Strategyproof approximation mechanisms for an obnoxious facility game on networks. *Theoretical Computer Science*, 497:154–163, 2013. DOI: 10.1016/j.tcs.2011.11.041. 61

W. J. Cho. Probabilistic assignment: A two-fold axiomatic approach. Mimeo, 2012. 26

G. Christodoulou and E. Koutsoupias. The price of anarchy of finite congestion games. In *Proc. of the 37th Annual ACM Symposium on the Theory of Computing (STOC)*, pages 67–73, 2005. DOI: 10.1145/1060590.1060600. 102

E. H. Clarke. Multipart pricing of public goods. *Public Choice*, 11:17–33, 1971. DOI: 10.1007/bf01726210. 49, 50

E. Clough. Strategic voting under conditions of uncertainty: A re-evaluation of Duverger's Law. *British Journal of Political Science*, 37(2):313–332, 2007. DOI: 10.1017/s0007123407000154. 87

V. Conitzer and T. Sandholm. Nonexistence of voting rules that are usually hard to manipulate. In *Proc. of the 21st Conference on Artificial Intelligence (AAAI)*, pages 627–634, 2006. 23

V. Conitzer, T. Sandholm, and J. Lang. When are elections with few candidates hard to manipulate? *Journal of the ACM*, 54(3):1–33, 2007. DOI: 10.1145/1236457.1236461. 37

V. Conitzer, T. Walsh, and L. Xia. Dominating manipulations in voting with partial information. In *Proc. of the 25th Conference on Artificial Intelligence (AAAI)*, 2011. 115

G. W. Cox. Duverger's law and strategic voting. unpublished paper, Department of Government, University of Texas, Austin, 1987. 82

G. W. Cox. Strategic voting equilibria under the single nontransferable vote. *American Political Science Review*, 88(3):608–621, 1994. DOI: 10.2307/2944798. 82

P. Cramton. The FCC spectrum auctions: An early assessment. *Journal of Economics and Management Strategy*, 6(3):431–495, 1997. DOI: 10.1111/j.1430-9134.1997.00431.x. 51

J. Davies, G. Katsirelos, N. Narodytska, and T. Walsh. Complexity of and algorithms for borda manipulation. In *Proc. of the 25th Conference on Artificial Intelligence (AAAI)*, pages 657–662, 2011. DOI: 10.1016/j.artint.2014.07.005. 38

Y. Desmedt and E. Elkind. Equilibria of plurality voting with abstentions. In *Proc. of the 11th Conference on Electronic Commerce (ACM-EC)*, pages 347–356, 2010. DOI: 10.1145/1807342.1807398. 79, 104

N. R. Devanur, M. Mihail, and V. V. Vazirani. Strategyproof cost-sharing mechanisms for set cover and facility location games. *Decision Support Systems*, 39(1):11–22, 2005. DOI: 10.1016/j.dss.2004.08.004. 61

A. Dhillon and B. Lockwood. When are plurality rule voting games dominance-solvable? *Games and Economic Behavior*, 46(1):55–75, 2004. DOI: 10.1016/s0899-8256(03)00050-2. 91

F. Dietrich and C. List. Arrow's theorem in judgment aggregation. *Social Choice and Welfare*, 29(1):19–33, 2007a. DOI: 10.1007/s00355-006-0196-x. 64

F. Dietrich and C. List. Strategyproof judgment aggregation. Open access publications from London School of Economics and Political Science, 2007b. 63, 64, 65

E. Dokow and D. Falik. Models of manipulation on aggregation of binary evaluations. *ArXiv Preprint ArXiv:1201.6388*, 2012. 64, 65

E. Dokow, M. Feldman, R. Meir, and I. Nehama. Mechanism design on discrete lines and cycles. In *Proc. of the 13th Conference on Electronic Commerce (ACM-EC)*, pages 423–440, 2012. DOI: 10.1145/2229012.2229045. 59, 61, 64

A. Downs. An economic theory of political action in a democracy. *Journal of Political Economy*, 65(2):135–150, 1957. DOI: 10.1086/257897. xv, 47, 54

J. Duggan and T. Schwartz. Strategic manipulability without resoluteness or shared beliefs: Gibbard-Satterthwaite generalized. *Social Choice and Welfare*, 17(1):85–93, 2000. DOI: 10.1007/pl00007177. 27

B. Dutta. Effectivity functions and acceptable game forms. *Econometrica: Journal of the Econometric Society*, 52(5):1151–1166, 1984. DOI: 10.2307/1910992. 11

B. Dutta and J.-F. Laslier. Costless honesty in voting. In *10th International Meeting of the Society for Social Choice and Welfare*, page 116, Moscow, 2010. 77

B. Dutta and A. Sen. Nash implementation with partially honest individuals. *Games and Economic Behavior*, 74(1):154–169, 2012. DOI: 10.1016/j.geb.2011.07.006. 78, 79

B. Dutta, H. Peters, and A. Sen. Strategyproof cardinal decision schemes. *Social Choice and Welfare*, 28(1):163–179, 2007. DOI: 10.1007/s00355-008-0299-7. 40

M. Duverger. *Political Parties: Their Organization and Activity in the Modern State.* North, B. and North, R., tr. New York, Wiley, Science Ed., 1963. 71

C. Dwork and J. Lei. Differential privacy and robust statistics. In *Proc. of the 41st Annual ACM Symposium on the Theory of Computing (STOC)*, pages 371–380, 2009. DOI: 10.1145/1536414.1536466. 43

C. Dwork, R. Kumar, M. Naor, and D. Sivakumar. Rank aggregation methods for the Web. In *Proc. of the 10th International Conference on World Wide Web (WWW)*, pages 613–622, 2001. DOI: 10.1145/371920.372165. 36

E. Elkind and A. Slinko. Rationalizations of voting rules. In F. Brandt, V. Conitzer, U. Endriss, J. Lang, and A. D. Procaccia, Eds., *Handbook of Computational Social Choice*, Cambridge University Press, 2016. DOI: 10.1017/cbo9781107446984. 6

E. Elkind, U. Grandi, F. Rossi, and A. Slinko. Gibbard-Satterthwaite games. In *IJCAI*, pages 533–539, 2015a. 26

E. Elkind, E. Markakis, S. Obraztsova, and P. Skowron. Equilibria of plurality voting: Lazy and truth-biased voters. In *Proc. of the 8th Symposium on Algorithmic Game Theory (SAGT)*, pages 110–122, 2015b. DOI: 10.1007/978-3-662-48433-3_9. 80

E. Elkind, E. Markakis, S. Obraztsova, and P. Skowron. Complexity of finding equilibria of plurality voting under structured preferences. In *Proc. of the 15th International Conference on Autonomous Agents and Multi-Agent Systems (AAMAS)*, pages 394–401, 2016. 80

U. Endriss. Judgment aggregation. In F. Brandt, V. Conitzer, U. Endriss, J. Lang, and A. D. Procaccia, Eds., *Handbook of Computational Social Choice*, Cambridge University Press, 2016. DOI: 10.1017/cbo9781107446984. 62, 63, 64

U. Endriss, U. Grandi, and D. Porello. Complexity of judgment aggregation. *Journal of Artificial Intelligence Research*, 45:481–514, 2012. DOI: 10.1613/jair.3708. 65

U. Endriss, S. Obraztsova, M. Polukarov, and J. S. Rosenschein. Strategic voting with incomplete information. In *Proc. of the 25th International Joint Conference on Artificial Intelligence (IJCAI)*, 2016. 115, 118

B. Escoffier, L. Gourves, N. Kim Thang, F. Pascual, and O. Spanjaard. Strategyproof mechanisms for facility location games with many facilities. *Algorithmic Decision Theory*, pages 67–81, 2011. DOI: 10.1007/978-3-642-24873-3_6. 61

A. Fabrikant, A. D. Jaggard, and M. Schapira. On the structure of weakly acyclic games. In *Proc. of the 3rd Symposium on Algorithmic Game Theory (SAGT)*, pages 126–137, 2010. DOI: 10.1007/978-3-642-16170-4_12. 101

D. Falik, R. Meir, and M. Tennenholtz. On coalitions and stable winners in plurality. In *Proc. of the 8th International Workshop on Internet and Network Economics (WINE)*, pages 256–269, 2012. DOI: 10.1007/978-3-642-35311-6_19. 77

P. Faliszewski and A. D. Procaccia. AI's war on manipulation: Are we winning? *AI Magazine*, 31(4):53–64, 2010. DOI: 10.1609/aimag.v31i4.2314. 38

J.-C. Falmagne and M. Regenwetter. A random utility model for approval voting. *Journal of Mathematical Psychology*, 40(2):152–159, 1996. DOI: 10.1006/jmps.1996.0014. 129

R. Farquharson. *Theory of Voting*, Yale University Press, 1969. 103

M. Feldman, A. Fiat, and I. Golomb. On voting and facility location. In *Proc. of the 17th ACM Conference on Electronic Commerce (ACM-EC)*, pages 269–286, 2016. DOI: 10.1145/2940716.2940725. 57, 58, 59

J. A. Ferejohn and M. P. Fiorina. The paradox of not voting: A decision theoretic analysis. *The American Political Science Review*, pages 525–536, 1974. DOI: 10.2307/1959502. 88, 89

M. Fey. Stability and coordination in Duverger's Law: A formal model of preelection polls and strategic voting. *American Political Science Review*, 91(1):135–147, 1997. DOI: 10.2307/2952264. 84, 85, 86

A. Filos-Ratsikas. *Social Welfare in Algorithmic Mechanism Design Without Money*, Ph.D. thesis, Department Office Computer Science, Aarhus University, 2015. 53

A. Filos-Ratsikas and P. B. Miltersen. Truthful approximations to range voting. In *Proc. of the 10th International Workshop on Internet and Network Economics (WINE)*, pages 175–188, 2014. DOI: 10.1007/978-3-319-13129-0_13. 52, 53

P. C. Fishburn and S. J. Brams. Paradoxes of preferential voting. *Mathematics Magazine*, 56(4):207–214, 1983. DOI: 10.2307/2689808. 79

S. D. Fisher. Definition and measurement of tactical voting: The role of rational choice. *British Journal of Political Science*, 34(1):152–166, 2004. DOI: 10.1017/s0007123403220391. 127

R. Forsythe, T. Rietz, R. Myerson, and R. Weber. An experimental study of voting rules and polls in three candidate elections. *International Journal of Game Theory*, 25(3):355–383, 1996. DOI: 10.1007/bf02425262. 128, 129

E. Friedgut, G. Kalai, and N. Nisan. Elections can be manipulated often. In *49th Annual IEEE Symposium on Foundations of Computer Science*, pages 243–249, 2008. DOI: 10.1109/focs.2008.87. 23

C. Geist and U. Endriss. Automated search for impossibility theorems in social choice theory: Ranking sets of objects. *Journal of Artificial Intelligence Research*, 40(1):143–174, 2011. 26

A. Gibbard. Manipulation of voting schemes: A general result. *Econometrica: Journal of the Econometric Society*, 41:587–601, 1973. DOI: 10.2307/1914083. 18

A. Gibbard. Manipulation of schemes that mix voting with chance. *Econometrica: Journal of the Econometric Society*, 45:665–681, 1977. DOI: 10.2307/1911681. 39, 41, 44, 52

L. Gourvès, J. Lesca, and A. Wilczynski. Strategic voting in a social context: Considerate equilibria. In *Proc. of the 22nd European Conference on Artificial Intelligence (ECAI)*, 2016. DOI: 10.3233/978-1-61499-672-9-1423. 101

U. Grandi, A. Loreggia, F. Rossi, K. B. Venable, and T. Walsh. Restricted manipulation in iterative voting: Condorcet efficiency and Borda score. In *Proc. of the 3rd International Conference on Algorithmic Decision Theory (ADT)*, pages 181–192, 2013. DOI: 10.1007/978-3-642-41575-3_14. 110, 116, 121, 123

R. J. Gretlein. Dominance solvable voting schemes: A comment. *Econometrica: Journal of the Econometric Society*, 50(2):527, 1982. DOI: 10.2307/1912643. 91

R. J. Gretlein. Dominance elimination procedures on finite alternative games. *International Journal of Game Theory*, 12(2):107–113, 1983. DOI: 10.1007/bf01774300. 91

T. Groves. Incentives in teams. *Econometrica: Journal of the Econometric Society*, 41(4):617–631, 1973. DOI: 10.2307/1914085. 49

J. Guiver and E. Snelson. Bayesian inference for Plackett-Luce ranking models. In *Proc. of the 26th International Conference on Machine Learning (ICML)*, pages 377–384, 2009. DOI: 10.1145/1553374.1553423. 120

M. Guo and V. Conitzer. Worst-case optimal redistribution of VCG payments in multi-unit auctions. *Games and Economic Behavior*, 67(1), pp. 69–98, 2009. 51

M. G. Hall. Electoral politics and strategic voting in state supreme courts. *The Journal of Politics*, 54(02):427–446, 1992. DOI: 10.2307/2132033. 127

J. D. Hartline. *Mechanism Design and Approximation*, 2013. Book draft. An updated version available from `http://jasonhartline.com/MDnA/` 48

J. C. Heckelman. Probabilistic borda rule voting. *Social Choice and Welfare*, 21(3):455–468, 2003. DOI: 10.1007/s00355-003-0211-4. 41, 42

R. Q. Herzberg and R. K. Wilson. Results on sophisticated voting in an experimental setting. *The Journal of Politics*, 50(2):471–486, 1988. DOI: 10.2307/2131804. 127

M. J. Hinich. Equilibrium in spatial voting: The median voter result is an artifact. *Journal of Economic Theory*, 16(2):208–219, 1977. DOI: 10.1016/0022-0531(77)90005-9. 54

R. Holzman. On strong representations of games by social choice functions. *Journal of Mathematical Economics*, 15(1):39–57, 1986. DOI: 10.1016/0304-4068(86)90022-4. 75

R. Holzman. Sub-core solutions of the problem of strong implementation. *International Journal of Game Theory*, 16(4):263–289, 1987. DOI: 10.1007/bf01756025. 75

A. Hylland. Strategy proofness of voting procedures with lotteries as outcomes and infinite sets of strategies. Unpublished paper, University of Oslo, 1980. 40

R. W. Jackman. Political institutions and voter turnout in the industrial democracies. *American Political Science Review*, 81(2):405–423, 1987. DOI: 10.2307/1961959. 129

M. O. Jackson. Implementation in undominated strategies: A look at bounded mechanisms. *The Review of Economic Studies*, 59(4):757–775, 1992. DOI: 10.2307/2297996. 75, 76

E. Kalai and E. Muller. Characterization of domains admitting nondictatorial social welfare functions and nonmanipulable voting procedures. *Journal of Economic Theory*, 16:457–469, 1977. DOI: 10.1016/0022-0531(77)90019-9. 31

R. M. Karp. *Reducibility Among Combinatorial Problems*, Springer, 1972. DOI: 10.1007/978-1-4684-2001-2_9. 35, 37

N. Kartik, O. Tercieux, and R. Holden. Simple mechanisms and preferences for honesty. *Games and Economic Behavior*, 83:284–290, 2014. DOI: 10.1016/j.geb.2013.11.011. 79

M. Kearns, S. Judd, J. Tan, and J. Wortman. Behavioral experiments on biased voting in networks. *Proc. of the National Academy of Sciences*, 106(5):1347–1352, 2009. DOI: 10.1073/pnas.0808147106. 128

J. S. Kelly. Strategyproofness and social choice functions without singlevaluedness. *Econometrica: Journal of the Econometric Society*, pages 439–446, 1977. DOI: 10.2307/1911220. 26

J. Kemeny. Mathematics without numbers. *Daedalus*, 88, 1959. 36

A. Koolyk, T. Strangway, O. Lev, and J. S. Rosenschein. Convergence and quality of iterative voting under non-scoring rules. In *Proc. of the 26th International Joint Conference on Artificial Intelligence (IJCAI)*, pages 273–279, 2017. DOI: 10.24963/ijcai.2017/39. 101, 103, 110, 118

N. S. Kukushkin. Acyclicity of improvements in finite game forms. *International Journal of Game Theory*, 40(1):147–177, 2011. DOI: 10.1007/s00182-010-0231-0. 94, 100, 101

S. P. Lalley and E. G. Weyl. Quadratic voting. *ArXiv Preprint ArXiv:1409.0264*, 2014. DOI: 10.2139/ssrn.2003531. 49

J. Lang and M. Slavkovik. Judgment aggregation rules and voting rules. In *International Conference on Algorithmic Decision Theory*, pages 230–243, Springer, 2013. DOI: 10.1007/978-3-642-41575-3_18. 62

J.-F. Laslier. The leader rule: A model of strategic approval voting in a large electorate. *Journal of Theoretical Politics*, 21(1):113–136, 2009. DOI: 10.1177/0951629808097286. 111, 116

J.-F. Laslier. Laboratory experiments on approval voting. In J.-F. Laslier and R. Sanver, Eds., *Handbook on Approval Voting*, pages 339–356, Springer, 2010a. DOI: 10.1007/978-3-642-02839-7. 128

J.-F. Laslier. In silico voting experiments. In J.-F. Laslier and R. Sanver, Eds., *Handbook on Approval Voting*, pages 311–335, Springer, 2010b. DOI: 10.1007/978-3-642-02839-7. 111, 120, 121

A. Lehtinen. Behavioral heterogeneity under approval and plurality voting. *Handbook on Approval Voting*, pages 285–310, 2010. DOI: 10.1007/978-3-642-02839-7_12. 120, 122

J. Leskovec, D. P. Huttenlocher, and J. M. Kleinberg. Governance in social media: A case study of the Wikipedia promotion process. In *Proc. of the 4th International AAAI Conference on Web and Social Media (ICWSM)*, 2010. 129

O. Lev. *Agent Modeling of Human Interaction: Stability, Dynamics and Cooperation*, Ph.D. thesis, The Hebrew University of Jerusalem, 2015. 101

O. Lev and J. S. Rosenschein. Convergence of iterative voting. In *Proc. of the 11th International Conference on Autonomous Agents and Multi-Agent Systems (AAMAS)*, pages 611–618, 2012. 97, 100, 101

O. Lev and J. S. Rosenschein. Convergence of iterative scoring rules. *Journal of Artificial Intelligence Research*, 57:573–591, 2016. 101

O. Lev, R. Meir, S. Obraztsova, and M. Polukarov. Heuristic voting as ordinal dominance strategies. In *Proc. of the 7th International Workshop on Computational Social Choice (COMSOC)*, 2018. 115

D. K. Levine and T. R. Palfrey. The paradox of voter participation? A laboratory study. *American Political Science Review*, 101(1):143–158, 2007. DOI: 10.1017/s0003055407070013. 129

K. Leyton-Brown and Y. Shoham. Essentials of game theory: A concise multidisciplinary introduction. *Synthesis Lectures on Artificial Intelligence and Machine Learning*, 2(1):1–88, 2008. DOI: 10.2200/s00108ed1v01y200802aim003. xvi, 8, 104

C. List and P. Pettit. Aggregating sets of judgments: An impossibility result. *Economics and Philosophy*, 18(1):89–110, 2002. DOI: 10.1017/s0266267102001098. 62

C. List and C. Puppe. Judgement aggregation: A survey. In P. Pattanaik, P. Anand, and C. Puppe, Eds., *The Handbook of Rational and Social Choice*, Oxford University Press, 2009. DOI: 10.1093/acprof:oso/9780199290420.001.0001. 62

T. Lu, P. Tang, A. D. Procaccia, and C. Boutilier. Bayesian vote manipulation: Optimal strategies and impact on welfare. In *Proc. of the Twenty-Eighth Conference on Uncertainty in Artificial Intelligence*, pages 543–553, AUAI Press, 2012. 119

H. Ma, R. Meir, and D. C. Parkes. Social choice for agents with general utilities. In *Proc. of the 25th International Joint Conference on Artificial Intelligence*, pages 345–351, AAAI Press, 2016. 51

H. Ma, R. Meir, and D. C. Parkes. Social choice with non quasi-linear utilities. In *Proc. of the 17th ACM Conference on Electronic Commerce (ACM-EC)*, 2018. To appear. 51

E. Maskin. Nash equilibrium and welfare optimality. *The Review of Economic Studies*, 66(1):23–38, 1999. DOI: 10.1111/1467-937x.00076. 73, 74

H. Matsushima. Role of honesty in full implementation. *Journal of Economic Theory*, 139(1):353–359, 2008. DOI: 10.1016/j.jet.2007.06.006. 77

R. D. McKelvey and R. G. Niemi. A multistage game representation of sophisticated voting for binary procedures. *Journal of Economic Theory*, 18(1):1–22, 1978. DOI: 10.1016/0022-0531(78)90039-x. 104

R. D. McKelvey and T. R. Palfrey. Quantal response equilibria for normal form games. *Games and Economic Behavior*, 10(1):6–38, 1995. DOI: 10.1006/game.1995.1023. 87

R. D. McKelvey and J. W. Patty. A theory of voting in large elections. *Games and Economic Behavior*, 57(1):155–180, 2006. DOI: 10.1016/j.geb.2006.05.003. 87

R. Meir. Plurality voting under uncertainty. In *Proc. of the 29th Conference on Artificial Intelligence (AAAI)*, pages 2103–2109, 2015. 101, 113, 114, 117, 118

R. Meir. Strong and weak acyclicity in iterative voting. In *Proc. of the 6th International Workshop on Computational Social Choice (COMSOC)*, 2016a. DOI: 10.1007/978-3-662-53354-3_15. 97

R. Meir. Strong and weak acyclicity in iterative voting. In *Proc. of the 9th Symposium on Algorithmic Game Theory (SAGT)*, pages 182–194, 2016b. A full version was included in the Proceedings of COMSOC'16. DOI: 10.1007/978-3-662-53354-3_15. 98, 101

R. Meir. Random tie-breaking with stochastic dominance. *CoRR*, abs/1609.01682, 2016c. 26

R. Meir. Iterative voting. In U. Endriss, Ed., *Trends in Computational Social Choice*, Chapter 4, pages 69–86, AI Access, 2017. 93, 98

R. Meir, A. D. Procaccia, and J. S. Rosenschein. Strategyproof classification under constant hypotheses: A tale of two functions. In *Proc. of the 23rd Conference on Artificial Intelligence (AAAI)*, pages 126–131, 2008. 60

R. Meir, M. Polukarov, J. S. Rosenschein, and N. Jennings. Convergence to equilibria of plurality voting. In *Proc. of the 24th Conference on Artificial Intelligence (AAAI)*, pages 823–828, 2010. 77, 99, 101

R. Meir, S. Almagor, A. Michaely, and J. S. Rosenschein. Tight bounds for strategyproof classification. In *Proc. of the 10th International Conference on Autonomous Agents and Multi-Agent Systems (AAMAS)*, pages 319–326, 2011. DOI: 10.1016/j.artint.2012.03.008. 58

R. Meir, A. D. Procaccia, and J. S. Rosenschein. Algorithms for strategyproof classification. *Artificial Intelligence*, 186:123–156, 2012. DOI: 10.1016/j.artint.2012.03.008. 55, 57, 60

R. Meir, O. Lev, and J. S. Rosenschein. A local-dominance theory of voting equilibria. In *Proc. of the 15th ACM Conference on Electronic Commerce (ACM-EC)*, pages 313–330, 2014. DOI: 10.1145/2600057.2602860. 69, 103, 112, 114, 117, 118, 121, 124

R. Meir, M. Polukarov, J. S. Rosenschein, and N. R. Jennings. Iterative voting and acyclic games. *Artificial Intelligence*, 252:100–122, 2017. DOI: 10.1016/j.artint.2017.08.002. 93, 96, 98, 100, 101

V. Merlin and J. Naeve. Implementation of social choice correspondences via demanding equilibria. In *Proc. of the International Conference on Logic, Game Theory and Social Choice (LGS)*, pages 264–280, 1999. 76

S. Merrill. Strategic decisions under one-stage multi-candidate voting systems. *Public Choice*, 36(1):115–134, 1981. DOI: 10.1007/bf00163774. 84, 111, 112

M. Messner and M. Polborn. Robust political equilibria under plurality and runoff rule. *IGIER Working Paper*, 2005. DOI: 10.2139/ssrn.838085. 89

C. Mezzetti and L. Renou. Implementation in mixed Nash equilibrium. *Journal of Economic Theory*, 147(6):2357–2375, 2012. DOI: 10.1016/j.jet.2012.09.004. 76

I. Milchtaich. Congestion games with player-specific payoff functions. *Games and Economic Behavior*, 13(1):111–124, 1996. DOI: 10.1006/game.1996.0027. 94

D. Monderer and L. S. Shapley. Potential games. *Games and Economic Behavior*, 14(1):124–143, 1996. DOI: 10.1006/game.1996.0044. 94

J. Moore and R. Repullo. Subgame perfect implementation. *Econometrica: Journal of the Econometric Society*, pages 1191–1220, 1988. DOI: 10.2307/1911364. 76

E. Mossel and M. Z. Rácz. A quantitative Gibbard-Satterthwaite theorem without neutrality. In *Proc. of the 44th Annual ACM Symposium on the Theory of Computing (STOC)*, pages 1041–1060, 2012. DOI: 10.1145/2213977.2214071. 23

E. Mossel, A. D. Procaccia, and M. Z. Rácz. A smooth transition from powerlessness to absolute power. *Journal of Artificial Intelligence Research*, pages 923–951, 2013. DOI: 10.1613/jair.4125. 24

H. Moulin. Dominance solvable voting schemes. *Econometrica: Journal of the Econometric Society*, pages 1337–1351, 1979. DOI: 10.2307/1914004. 90, 91

H. Moulin. On strategyproofness and single-peakedness. *Public Choice*, 35:437–455, 1980. DOI: 10.1007/bf00128122. 31

H. Moulin. Condorcet's principle implies the no show paradox. *Journal of Economic Theory*, 45(1):53–64, 1988. DOI: 10.1016/0022-0531(88)90253-0. 79

H. Moulin. One-dimensional mechanism design. *Theoretical Economics*, 12(2):587–619, 2017. DOI: 10.3982/te2307. 61

H. Moulin and B. Peleg. Cores of effectivity functions and implementation theory. *Journal of Mathematical Economics*, 10(1):115–145, 1982. DOI: 10.1016/0304-4068(82)90009-x. 75

E. Muller and M. A. Satterthwaite. The equivalence of strong positive association and strategyproofness. *Journal of Economic Theory*, 14(2):412–418, 1977. DOI: 10.1016/0022-0531(77)90140-5. 18, 19

D. P. Myatt. On the theory of strategic voting. *The Review of Economic Studies*, 74(1):255–281, 2007. DOI: 10.1111/j.1467-937x.2007.00421.x. 85

R. B. Myerson and R. J. Weber. A theory of voting equilibria. *The American Political Science Review*, 87(1):102–114, 1993. DOI: 10.2307/2938959. 84, 85, 112

R. Nadeau, E. Cloutier, and J.-H. Guay. New evidence about the existence of a bandwagon effect in the opinion formation process. *International Political Science Review*, 14(2):203–213, 1993. DOI: 10.1177/019251219301400204. 129

K. Nehring and C. Puppe. The structure of strategy-proof social choice–part I: General characterization and possibility results on median spaces. *Journal of Economic Theory*, 135(1):269–305, 2007. DOI: 10.1016/j.jet.2006.04.008. 31, 33, 34

K. Nehring and C. Puppe. Abstract Arrowian aggregation. *Journal of Economic Theory*, 145(2):467–494, 2010. DOI: 10.1016/j.jet.2010.01.010. 65

N. Nisan. Introduction to mechanism design (for computer scientists). In N. Nisan, T. Roughgarden, E. Tardos, and V. Vazirani, Eds., *Algorithmic Game Theory*, Chapter 9, Cambridge University Press, 2007. DOI: 10.1017/cbo9780511800481. 47, 49

M. Nunez and J.-F. Laslier. Bargaining through approval. *Journal of Mathematical Economics*, 60:63–73, 2015. DOI: 10.1016/j.jmateco.2015.06.015. 79

M. Núnez and M. Pivato. Truth-revealing voting rules for large populations. In *Proc. of the 6th International Workshop on Computational Social Choice (COMSOC)*, 2016. DOI: 10.2139/ssrn.2802679. 42

S. Obraztsova, E. Markakis, and D. R. M. Thompson. Plurality voting with truth-biased agents. In *Proc. of the 6th Symposium on Algorithmic Game Theory (SAGT)*, pages 26–37, 2013. DOI: 10.1007/978-3-642-41392-6_3. 77

S. Obraztsova, E. Markakis, M. Polukarov, Z. Rabinovich, and N. R. Jennings. On the convergence of iterative voting: How restrictive should restricted dynamics be? In *Proc. of the 29th Conference on Artificial Intelligence (AAAI)*, pages 993–999, 2015. 110, 116

S. Obraztsova, O. Lev, M. Polukarov, Z. Rabinovich, and J. S. Rosenschein. Non-myopic voting dynamics: An optimistic approach. In *Proc. of the 10th Multidisciplinary Workshop on Advances in Preference Handling (M-PREF)*, 2016a. 113, 115, 118

S. Obraztsova, Z. Rabinovich, E. Elkind, M. Polukarov, and N. Jennings. Trembling hand equilibria of plurality voting. In *Proc. of the 25th International Joint Conference on Artificial Intelligence (IJCAI)*, 2016b. 90

S. Obraztsova, O. Lev, E. Markakis, Z. Rabinovich, and J. S. Rosenschein. Distant truth: Bias under vote distortion costs. In *Proc. of the 16th International Conference on Autonomous Agents and Multi-Agent Systems (AAMAS)*, pages 885–892, 2017. 78

M. J. Osborne and A. Rubinstein. Sampling equilibrium, with an application to strategic voting. *Games and Economic Behavior*, 45(2):434–441, 2003. DOI: 10.1016/s0899-8256(03)00147-7. 118

G. Owen and B. Grofman. To vote or not to vote: The paradox of nonvoting. *Public Choice*, 42(3):311–325, 1984. DOI: 10.1007/bf00124949. 79

T. R. Palfrey. *A Mathematical Proof of Duverger's Law*, 1988. 82, 84, 87

T. R. Palfrey. Implementation theory. *Handbook of Game Theory with Economic Applications*, 3:2271–2326, 2002. 72

T. R. Palfrey and H. Rosenthal. Testing game-theoretic models of free riding: New evidence on probability bias and learning. Mimeo, 1990. 86

T. R. Palfrey and S. Srivastava. Nash implementation using undominated strategies. *Econometrica: Journal of the Econometric Society*, pages 479–501, 1991. DOI: 10.2307/2938266. 76

B. Peleg and A. D. Procaccia. Implementation by mediated equilibrium. *International Journal of Game Theory*, 39(1–2):191–207, 2010. DOI: 10.1007/s00182-009-0175-4. 76

A. Postlewaite and D. Schmeidler. Strategic behaviour and a notion of ex ante efficiency in a voting model. *Social Choice and Welfare*, 3(1):37–49, 1986. DOI: 10.1007/bf00433523. 40

G. B. Powell. American voter turnout in comparative perspective. *American Political Science Review*, 80(1):17–43, 1986. DOI: 10.2307/1957082. 129

J. W. Pratt. Risk aversion in the small and in the large. *Econometrica: Journal of the Econometric Society*, pages 122–136, 1964. DOI: 10.2307/1912743. 51

A. D. Procaccia. Can approximation circumvent Gibbard-Satterthwaite? In *Proc. of the 24th Conference on Artificial Intelligence (AAAI)*, 2010. 41, 42

A. D. Procaccia and J. S. Rosenschein. Junta distributions and the average-case complexity of manipulating elections. *Journal of Artificial Intelligence Research*, 28:157–181, 2007. DOI: 10.1145/1160633.1160726. 23

A. D. Procaccia and M. Tennenholtz. Approximate mechanism design without money. In *Proc. of the 10th ACM Conference on Electronic Commerce (ACM-EC)*, pages 177–186, 2009. DOI: 10.1145/1566374.1566401. 54, 58

A. D. Procaccia and M. Tennenholtz. Approximate mechanism design without money. *ACM Transactions on Economics and Computation*, 1(4):18, 2013. DOI: 10.1145/2542174.2542175. 54

Z. Rabinovich, S. Obraztsova, O. Lev, E. Markakis, and J. S. Rosenschein. Analysis of equilibria in iterative voting schemes. In *Proc. of the 29th Conference on Artificial Intelligence (AAAI)*, pages 1007–1013, 2015. 100

S. R. Reed. Structure and behaviour: Extending Duverger's Law to the Japanese case. *British Journal of Political Science*, 20(3):335–356, 1990. DOI: 10.1017/s0007123400005871. 127

M. Regenwetter. *Behavioral Social Choice: Probabilistic Models, Statistical Inference, and Applications*, Cambridge University Press, 2006. 81

M. Regenwetter and B. Grofman. Approval voting, borda winners, and condorcet winners: Evidence from seven elections. *Management Science*, 44(4):520–533, 1998. DOI: 10.1287/mnsc.44.4.520. 128

A. Reijngoud and U. Endriss. Voter response to iterated poll information. In *Proc. of the 11th International Conference on Autonomous Agents and Multi-Agent Systems (AAMAS)*, pages 635–644, 2012. 109, 115, 116

K. Reinecke, M. K. Nguyen, A. Bernstein, M. Näf, and K. Z. Gajos. Doodle around the world: Online scheduling behavior reflects cultural differences in time perception and group decision-making. In *Proc. of the 16th Conference on Computer Supported Cooperative Work (CSCW)*, pages 45–54, 2013. DOI: 10.1145/2441776.2441784. 129

R. Reyhani and M. C. Wilson. Best-reply dynamics for scoring rules. In *Proc. of the 20th European Conference on Artificial Intelligence (ECAI)*, 2012. DOI: 10.3233/978-1-61499-098-7-672. 98, 99, 100, 101

R. Reyhani, M. C. Wilson, and J. Khazaei. Coordination via polling in plurality voting games under inertia. In *Proc. of the 4th International Workshop on Computational Social Choice (COMSOC)*, 2012. 115

H. W. Riecken. Primary groups and political party choice. In E. Burdick and A. Brodbeck, Eds., *American Voting Behavior*, The Free Press, 1959. DOI: 10.2307/1891594. 129

W. H. Riker. The number of political parties: A reexamination of Duverger's Law. *Comparative Politics*, 9(1):93–106, 1976. DOI: 10.2307/421293. 82

W. H. Riker. The two-party system and Duverger's law: An essay on the history of political science. *American Political Science Review*, 76(04):753–766, 1982. DOI: 10.2307/1962968. 71

W. H. Riker and P. C. Ordeshook. A theory of the calculus of voting. *American Political Science Review*, 62(1):25–42, 1968. DOI: 10.1017/s000305540011562x. 81

K. Roberts. The characterization of implementable choice rules. *Aggregation and Revelation of Preferences*, 12(2):321–348, 1979. 51

R. Rosenthal. A class of games possessing pure-strategy Nash equilibria. *International Journal of Game Theory*, 2:65–67, 1973. DOI: 10.1007/bf01737559. 70

M. H. Rothkopf. Thirteen reasons why the Vickrey-Clarke-Groves process is not practical. *Operations Research*, 55(2):191–197, 2007. DOI: 10.1287/opre.1070.0384. 48

M. Satterthwaite. Strategy-proofness and arrow's conditions: Existence and correspondence theorems for voting procedures and social welfare functions. *Journal of Economic Theory*, 10:187–216, 1975. DOI: 10.1016/0022-0531(75)90050-2. 18

L. J. Savage. The theory of statistical decision. *Journal of the American Statistical Association*, 46(253):55–67, 1951. DOI: 10.2307/2280094. 88

J. Schummer and R. V. Vohra. Strategyproof location on a network. *Journal of Economic Theory*, 104(2):405–428, 2004. DOI: 10.1006/jeth.2001.2807. 59, 60, 61

R. Selten. Reexamination of the perfectness concept for equilibrium points in extensive games. *International Journal of Game Theory*, 4(1):25–55, 1975. DOI: 10.1007/bf01766400. 90

M. R. Sertel and M. R. Sanver. Strong equilibrium outcomes of voting games are the generalized condorcet winners. *Social Choice and Welfare*, 22:331–347, 2004. DOI: 10.1007/s00355-003-0218-x. 74

S. Sina, N. Hazon, A. Hassidim, and S. Kraus. Adapting the social network to affect elections. In *Proc. of the 14th International Conference on Autonomous Agents and Multi-Agent Systems (AAMAS)*, pages 705–713, 2015. 87

T. Sjöström. Implementation in undominated Nash equilibria without integer games. *Games and Economic Behavior*, 6(3):502–511, 1994. DOI: 10.1006/game.1994.1029. 76

A. Slinko and S. White. Is it ever safe to vote strategically? *Social Choice and Welfare*, 43(2):403–427, 2014. DOI: 10.1007/s00355-013-0785-4. 24, 25

L.-G. Svensson. The proof of the Gibbard-Satterthwaite theorem revisited. Working Paper No. 1999:1, Department of Economics, Lund University, 1999. `http://www.nek.lu.se/NEKlgs/vote09.pdf` DOI: 10.1016/j.jmateco.2014.09.007. 18, 19, 20

M. Tal, R. Meir, and Y. Gal. A study of human behavior in voting systems. In *Proc. of the 14th International Conference on Autonomous Agents and Multi-Agent Systems (AAMAS)*, pages 665–673, 2015. 127, 128, 129

A. D. Taylor. The manipulability of voting systems. *The American Mathematical Monthly*, 109(4):321–337, 2002. DOI: 10.2307/2695497. 28

D. R. M. Thompson, O. Lev, K. Leyton-Brown, and J. S. Rosenschein. Empirical analysis of plurality election equilibria. In *Proc. of the 12th International Conference on Autonomous Agents and Multi-Agent Systems (AAMAS)*, pages 391–398, 2013. 78

A. Tsang and K. Larson. The echo chamber: Strategic voting and homophily in social networks. In *Proc. of the 15th International Conference on Autonomous Agents and Multi-Agent Systems (AAMAS)*, pages 368–375, 2016. 87, 121

M. Tyszler and A. Schram. Information and strategic voting. *Experimental Economics*, 19(2):360–381, 2016. DOI: 10.1007/s10683-015-9443-2. 127

A. M. van Deemen and N. P. Vergunst. Empirical evidence of paradoxes of voting in Dutch elections. *Public Choice*, pages 475–490, 1998. DOI: 10.1007/978-1-4757-5127-7_11. 127

K. Van der Straeten, J.-F. Laslier, N. Sauger, and A. Blais. Strategic, sincere, and heuristic voting under four election rules: An experimental study. *Social Choice and Welfare*, 35(3):435–472, 2010. DOI: 10.1007/s00355-010-0448-7. 127, 128

H. van Ditmarsch, J. Lang, and A. Saffidine. Strategic voting and the logic of knowledge. In *Proc. of the 14th Conference on Theoretical Aspects of Rationality and Knowledge (TARK)*, 2013. 115

J. Van Leeuwen. *Handbook of Theoretical Computer Science (Vol. A): Algorithms and Complexity*, MIT Press, 1991. 35

W. Vickrey. Counter speculation, auctions, and competitive sealed tenders. *The Journal of Finance*, 16(1):8–37, 1961. DOI: 10.2307/2977633. 49

M. Vorsatz. Approval voting on dichotomous preferences. *Social Choice and Welfare*, 28(1):127–141, 2007. DOI: 10.1007/s00355-006-0149-4. 35

L. Xia and V. Conitzer. A sufficient condition for voting rules to be frequently manipulable. In *Proc. of the 9th ACM Conference on Electronic Commerce (ACM-EC)*, pages 99–108, 2008. DOI: 10.1145/1386790.1386810. 23, 24

L. Xia, V. Conitzer, and J. Lang. Strategic sequential voting in multi-issue domains and multiple-election paradoxes. In *Proc. of the 12th ACM Conference on Electronic Commerce (ACM-EC)*, pages 179–188, 2011. DOI: 10.1145/1993574.1993602. 71, 104

H. P. Young. The evolution of conventions. *Econometrica: Journal of the Econometric Society*, 61(1):57–84, 1993. DOI: 10.2307/2951778. 94

J. Zou, R. Meir, and D. Parkes. Strategic voting behavior in doodle polls. In *Proc. of the 18th Conference on Computer Supported Cooperative Work (CSCW)*, pages 464–472, 2015. DOI: 10.1145/2675133.2675273. 129

W. S. Zwicker. Introduction to the theory of voting. In F. Brandt, V. Conitzer, U. Endriss, J. Lang, and A. D. Procaccia, Eds., *Handbook of Computational Social Choice*, Cambridge University Press, 2016. DOI: 10.1017/cbo9781107446984. xvi, 5

Author's Biography

RESHEF MEIR

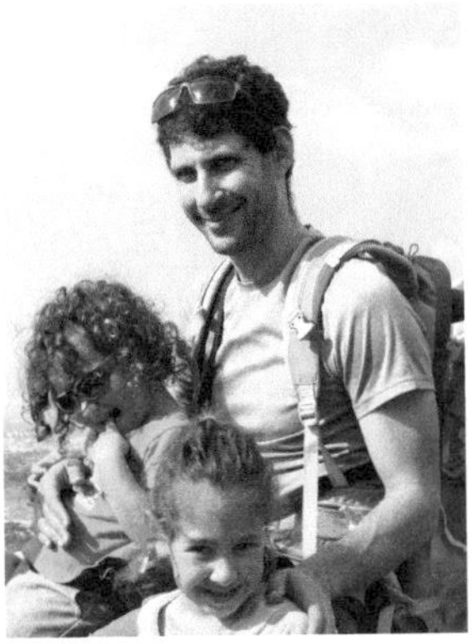

Reshef Meir is a senior lecturer in the Department of Industrial Engineering at Technion–Israel Institute of Technology in Haifa, Israel. Prior to that, he completed a postdoctorate at the Harvard Center for Research on Computation and Society and a Ph.D. in computer science from the Hebrew University. His main research areas are computational game theory, mechanism design, social choice, and bounded rationality.

Dr. Meir was named one of AI's Ten to Watch by *IEEE Intelligent Systems* (2015). In addition, he received the Michael B. Maschler Prize (Game Theory Society), a Rothschild postdoctoral fellowship, the Alon fellowship for young faculty, and an honorable mention for Victor Lesser Distinguished Dissertation Award by IFAAMAS.

Printed in the United States
by Baker & Taylor Publisher Services